United States and Mexico

Ties That Bind, Issues That Divide

Emma Aguila, Alisher R. Akhmedjonov, Ricardo Basurto-Davila, Krishna B. Kumar, Sarah Kups, Howard J. Shatz

INVESTMENT IN PEOPLE AND IDEAS

This monograph results from the RAND Corporation's Investment in People and Ideas program. Support for this program is provided, in part, by the generosity of RAND's donors and by the fees earned on client-funded research.

The authors of this monograph are listed in alphabetical order.

Library of Congress Control Number: 2012937838

ISBN: 978-0-8330-5106-6

Published 2012 by the RAND Corporation
1776 Main Street, P.O. Box 2138, Santa Monica, CA 90407-2138
1200 South Hayes Street, Arlington, VA 22202-5050
4570 Fifth Avenue, Suite 600, Pittsburgh, PA 15213-2665
RAND URL: http://www.rand.org/
To order RAND documents or to obtain additional information, contact
Distribution Services: Telephone: (310) 451-7002;
Fax: (310) 451-6915; Email: order@rand.org

Preface

Despite geographical closeness and many shared economic interests, wariness has characterized relations between the United States and Mexico during the past few decades. Policies designed to curtail the number of Mexican immigrants entering the United States, a 700-mile-long border fence between the two countries, and an increasing illegal drug trade have somewhat eclipsed the North American Free Trade Agreement's (NAFTA's) cooperative scope. However, to ensure that the economic and political interaction between the two countries is as mutually beneficial as it is sustainable, it is critical that Mexico and the United States reiterate their commitment to their important relationship. This monograph focuses on how the alliance between the United States and Mexico can be made stronger.

We undertook answering this question by drawing on the perspectives of both countries. This multidisciplinary, binational approach combines economics, demography, and sociology, as well as discussions with U.S. and Mexican policymakers and reviews of published work and of survey results, in order to present a comprehensive analysis of issues that concern each country. The first part of the monograph is a detailed discussion of the issues surrounding migration from Mexico from a U.S. perspective. The second part of the monograph presents the economic and social challenges faced by Mexico. The third part reframes questions of immigration and trade pertinent to both countries by examining the results of opinion polls and trade policy. The final part presents our conclusions and recommendations.

This monograph should be of interest to Mexican and U.S. policymakers, to foreign policy experts, and to organizations and individuals committed to helping to sustain a successful alliance between NAFTA countries. The monograph is a companion to another RAND monograph that examines the security situation in Mexico and assesses its impact on the United States.

This monograph results from the RAND Corporation's Investment in People and Ideas program. Support for this program is provided, in part, by the generosity of RAND's donors and by the fees earned on client-funded research.

The authors of this report are listed in alphabetical order. Questions or comments about this report are welcome and should be directed to the project leaders:

Emma Aguila
RAND Corporation
1776 Main Street
Santa Monica, California 90401-3208
310-393-0411 x6682
Emma_Aguila@rand.org

Krishna B. Kumar
RAND Corporation
1776 Main Street
Santa Monica, California 90401-3208
310-393-0411 x7589
Krishna_Kumar@rand.org

More information about RAND is available at http://www.rand.org.

Contents

Figures

Tables

Summary

The purpose of this study was to assess ways to strengthen the alliance between the United States and Mexico. In this monograph, we provide objective analysis of issues relevant to this unique international relationship. It is our belief that both Mexican and U.S. policymakers would benefit from a detailed discussion of immigration and the social and economic development of Mexico, interrelated issues connecting both countries. Understanding the concerns that drive the debate on immigration in the United States might better enable Mexican policymakers to engage in a constructive dialogue that helps support and reform U.S. immigration policy. In addition, U.S. policymakers might benefit from understanding the social and economic achievements of, as well as challenges faced by, Mexico. Such factors influence many Mexicans' decisions to emigrate across the border every year. Although we recognize that there are other issues of mutual interest to both countries, immigration to the United States and the social and economic backdrop of Mexico are topics complex enough to illustrate the progress made by and the challenges that remain for both countries.[1]

The United States and Mexico have always shared a complicated bilateral relationship, which is clearly evidenced by the heavy migration flows and trade agreements of the past 20 years. A major issue of contention between both countries is that of immigration from Mexico to the United States. The September 11 attacks promulgated a tightening of immigration restrictions, which led to the Secure Fence Act of

[1] The issue of security in Mexico is dealt with in a companion piece to this monograph (Schaefer, Bahney, and Riley, 2009).

2006 (Pub. L. 109-367), supporting the construction of a 700-mile-long fence along the Mexico–United States border. On the other hand, with the elimination of tariffs and other economic trade barriers in the mid-1990s, the North American Free Trade Agreement (NAFTA) has served as a multilateral platform for cooperation between both countries, as well as Canada. In March 2011, U.S. President Barack Obama met with Mexican President Felipe Calderón to discuss working together on issues ranging from drug-related violence, immigration reform, and cross-border trucking to climate change and international politics.

Objectives and Approach of the Monograph

The project had three objectives. First, we described the conditions precipitating the large immigration flow of Mexican citizens to the United States, especially during the 1990s and early part of this century. Second, we examined specific social reforms and challenges faced by Mexican citizens in the period before the U.S. recession that started in 2007 and the global crisis that commenced in 2008. Although detailed statistical and economic analyses of the recession continue to be produced, our aim is to analyze secular trends over the past few decades that relate to structural issues. Toward this end, we do make note of data available since 2007 that highlight noteworthy patterns that hold true across *previous* economic crises to understand their impact on Mexico's socioeconomic development. Finally, we studied the prevailing opinions that U.S. citizens have of Mexican immigration.

Our multidisciplinary approach entailed the following:

- reviewing and documenting contemporary and historical policy contexts of Mexican immigration to the United States
- collecting and analyzing information on the economic conditions, social structures, and government programs in Mexico
- interviewing U.S. and Mexican policymakers about factors influencing Mexican migration and U.S.-Mexican relations

- describing U.S.-Mexican policy up to 2011, and assessing popular public opinion on U.S.-Mexican relations and migration.

To the best of our knowledge, the information presented is the most recent available as of March 2011, by which date the data for this monograph were gathered, but we must caution that the subject matter concerns a fast-moving policy area, and there are constant changes. The manuscript is designed as a binational reference for both U.S. and Mexican policymakers. We define *binational* as a way of examining an issue while taking into account the concerns and interests of both Mexico and the United States. More clearly, the monograph presents Mexican immigration to the United States and the economic and social development of Mexico as interrelated issues. Differences in economic growth, wages, and the employment situation between the two countries are critical determinants of immigration. Migration of labor out of Mexico, in turn, affects Mexico's economic and social development. Policymakers of both countries should be cognizant of the connections between immigration and economic and social development in order to design policies that address these complex issues. Moreover, because Mexico is an important economic partner of the United States, a better comprehension of the barriers and challenges to Mexico's economic growth can provide U.S. decisionmakers with insights into determining which future policies might be beneficial to both countries.

The Mexican Migration Situation

During the 20th and 21st centuries, the United States and Mexico have simultaneously promoted and discouraged migration from Mexico to the United States through their migration policies. Although the United States has mostly attempted to limit the volume of Mexican immigration, it actively promoted the immigration of Mexican workers in response to labor shortages during World Wars I and II. Similarly, Mexico has, at different times, encouraged labor migration to the United States, tried to control the number of emigrants, or instead simply taken a laissez-faire approach to the flow of illegal migrants through its northern border.

Our review of the research literature, immigration policy documents, and statistical and economic data yielded the composite picture, described next, of the state of Mexican immigration to the United States.

The Number of Mexican Immigrants in the United States Increased Significantly in the Past 20 Years

By 2005, an estimated 16 percent of Mexico's working-age population resided in the United States (Giorguli Saucedo, Olvera, and Leite, 2006). The size of the Mexican population living in the United States has increased significantly since the 1990s and has been estimated at 11.5 million in 2009 (Pew Hispanic Center, 2009) (Figure S.1). Migrants have also expanded their destinations in the United States from traditional migrant-receiving states—Arizona, California, Florida, Illinois, and Texas (see Figure S.2)—to the entire country. As of 2009, Mexicans made up the largest immigrant group in 33 states, and

Figure S.1
Mexican Immigrants in the United States, 1960–2009

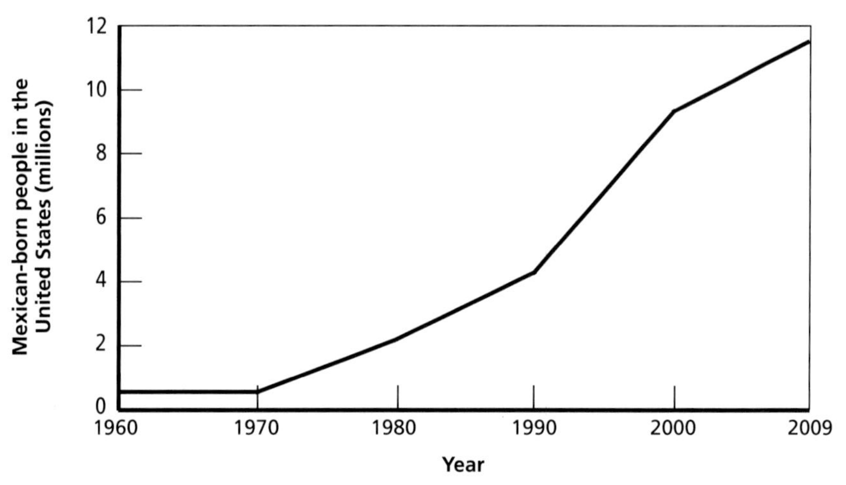

SOURCE: Migration Policy Institute, undated.
RAND *MG985/1-S.1*

in the home country, including lack of jobs, economic crises, poverty, and crop failure. Finally, social networks play a role in decisions to emigrate. U.S. employers frequently seek additional low-wage Mexican workers through current immigrant employees. Acquaintances living in the United States can ease the migratory process for new migrants.

An Important Concern in the United States Regarding Immigration Is Its Economic Impact

Although the evidence is inconclusive and the debate is still ongoing, results indicate that, between 1980 and 2000, the wages of low-skilled U.S. workers, who represent about 8 percent of the native labor force (Hanson, 2009), fell by up to 9 percent due to all immigration (Borjas, 2003). However, other studies suggest that immigration increases economic activity by attracting more businesses and thus creating more jobs (see, e.g., Peri, 2009). In addition, immigration might also have positive economic effects by lowering the prices of labor-intensive goods and services (Cortes, 2008). In terms of fiscal burden—whether immigrants receive more in government benefits than they pay in taxes—it appears that the net fiscal impact of immigration is small, although much of this burden is likely borne by state governments (Smith and Edmonston, 1997).

In the academic literature, there is some evidence that return migration flows are driven by the economic conditions in the country of origin rather than in the country of destination (Papademetriou and Terrazas, 2009). However, in the media (see Alarcón, 2008; Camarota and Jensenius, 2009), there has been an expectation of a sharp increase in return migration as a result of the economic crisis the United States faced starting in 2007. However, data show low levels of return migration between 2007 and 2009 (Rendall, Brownell, and Kups, 2010). As of early 2011, Mexico was recovering from a deep economic downturn as a result of the global economic crises that were triggered by the 2007–2009 financial and economic crises in the United States, as well as the H1N1 outbreak in the spring of 2009. This will more likely encourage the trend of immigrants staying for longer periods, thereby lowering the return flow of migrants.

The Economic and Social Situation in Mexico

In the mid-1990s, Mexico suffered a deep economic crisis. NAFTA and economic reforms aimed at increasing competitiveness created expectations of sustainable economic growth. However, a sudden depreciation of the Mexican peso in December 1994 started a financial crisis that resulted in a 6.2-percent contraction in gross domestic product (GDP) between 1994 and 1995 and a rise in inflation, which peaked at 35 percent. Recovery from the crisis took approximately five years, but the period that followed the recovery up to 2008 was characterized by macroeconomic stability, including steady exchange and interest rates and low inflation. Nevertheless, GDP per capita and productivity growth during those years were modest.

Mexico's economic performance has been relatively stable from the late 1990s to the late 2000s. Given divided governments and the slow democratic process, its reforms have been impressive. But more reforms are needed, especially in the areas of fiscal, labor, and energy policy. The rapid growth of trade between the United States and Mexico seen after NAFTA has slowed since 2000, and challenges remain.

Our interviews with Mexican officials, study of Mexican social and economic policy documents, and statistical and financial data produced the following composite picture of Mexico's social and economic conditions:

- *Economic competitiveness with other countries remains mixed.* Despite the multiple economic reforms of the 1990s, Mexico long remained behind other developing countries in terms of competitiveness but has been catching up in some categories. In the 2011 *Global Competitiveness Index* (GCI) by the World Economic Forum, Mexico ranks 58th worldwide. China, India, and Chile are some of the countries that compete with Mexico for foreign investment, and these countries had better placement in these rankings. However, in the World Bank's *Doing Business* rankings, Mexico ranks 35th and higher than these three countries (World Bank, undated [b]). The problems commonly identified with Mexico are corruption, a weak judicial system, and excessive bureaucracy. It has been estimated that regulatory burden costs Mexico 15 percent of its GDP. Also related to competitiveness is the need for better labor-market policies; labor regulations in Mexico are among the most rigid in the Organisation for Economic Co-Operation and Development (OECD) and emerging markets, a fact that adds considerable costs to employment, stimulates informal markets, and deters entrepreneurship.

- *Taxation revenues are low.* Fiscal policy is another area in which there is still much to be done. Taxes as a percentage of GDP are low not only compared with those of other Latin American countries but also compared with those of the United States and the OECD average. In addition, the Mexican government relies heavily on income from Petróleos Mexicanos (PEMEX), the state-owned oil company; about one-third of public-sector revenues are oil related. An important reason for low tax revenues and high reliance on oil taxes is the vast tax evasion in the informal economy, which is estimated to account for 20 percent of the profits generated in the country. Fiscal reforms enacted in 2007 are expected to increase nonoil tax revenue by 2.8 percent of the GDP by 2012, but further efforts are needed to expand the tax base and avoid tax evasion.
- *Energy development is becoming one of Mexico's most important challenges for the future.* Oil production is projected to decline such that Mexico, the seventh-largest oil producer in the world as of 2010, could become a net oil importer by 2018. Preventing this situation will require heavy investment in exploration and development of oil extraction methods, but this is unattainable because PEMEX lacks the needed financial resources because of its transfers to the federal government, which have resulted in its being one of the most heavily indebted oil company in the world. In addition, regulations severely limit PEMEX's ability to make strategic decisions or engage in partnerships that could provide it with the required technology. A reform initiative passed by the Mexican Congress in October 2008 represents a small step in the right direction, allowing PEMEX to subcontract foreign compa-

nies for exploration and drilling, but further regulatory efforts with more far-reaching changes are still needed.

- *Money sent by Mexican immigrants in the United States to their families in Mexico has become a major source of income.* An important determinant of Mexican social conditions is the inflow of remittances. From 2000 to 2009, remittances were the second-highest source of external resources for Mexico, only slightly below foreign direct investment (FDI). International organizations have recently shown interest in the role of remittances as a tool for economic development. Federal and state governments in Mexico have created programs to channel remittances sent by migrant organizations toward financing infrastructure, public services, and other community-related projects. This makes Mexico vulnerable to changes due to fluctuations in the U.S. economy.
- *Mexico depends on the United States for foreign trade and investment.* Aided by the enactment of NAFTA in 1994, Mexico has become the third-largest trading partner for the United States, behind only Canada and China. Relative FDI levels have been lower: In 2010, U.S. FDI in Mexico amounted to only 2.3 percent of worldwide U.S. FDI, measured as the U.S. direct investment position on a historical cost basis, similar to the 1994 percentage, although it did rise and then fall (BEA, 2010; Lowe, 2010). However, Mexico's trade and investment dependence on the United States is far higher. In 2010, more than 79.9 percent of Mexican merchandise exports were destined for the United States, and 48.1 percent of Mexico's merchandise imports came from the United States. In that same year, 27.6 percent of Mexico's FDI came from the United States, but that was an unusual year. That figure averaged 53.9 percent between 1994 and 2010 and was above 40 percent every year from 2005 to 2009.
- *Poverty in Mexico is still widespread.* Mexico's progress in alleviating poverty has been mixed in the past two decades. Depending on the measure used, economic inequality has either remained unchanged or grown, but it certainly has not decreased. Government figures show an important reduction in poverty rates between 1996 and 2006. However, between 2006 and 2010, this

was somewhat reversed, and poverty is still widespread, particularly in rural areas and among indigenous communities, and increased during the economic crisis. Mexico has a long tradition of instituting social programs, particularly those with the goal of alleviating poverty. Although evaluations of these programs are rare, Oportunidades—the largest program focusing on poverty alleviation—was designed with a rigorous and independent evaluation system. This program provides cash transfers to households in extreme poverty, conditional on those households fulfilling certain obligations, including keeping their children in school and attending clinics for health education and medical exams. Evaluations of Oportunidades have highlighted positive results in several areas, and several governments of other countries have begun implementing similar programs. There are still many challenges in this area, particularly in the less developed regions in Mexico.

- *The quality of education in Mexico is low.* Mexican students generally obtain low scores in international studies of academic achievement, and grade repetition and dropout rates are relatively high. Some progress has been made in access to basic education, but access to higher education is still low; only 8 percent of Mexico's population hold bachelor's degrees. Almost 80 percent of the education budget is allocated to teacher compensation, leaving little left to invest in other educational resources.
- *Mexico has shown improvement in providing health care for its citizens during the past half-century.* Life expectancy increased by almost 28 years between 1950 and 2010, and infant mortality declined 64 percent between 1990 and 2010. An important shift in the health-care system is taking place as Mexico undergoes an epidemiological transition from infectious diseases to degenerative conditions, such as diabetes and cardiovascular disease. However, this transition is not occurring uniformly across the country: The risk of death due to transmissible diseases or malnutrition is 30 percent higher in rural communities than in urban areas. Another cause of concern is the relatively low level of resources being allocated to the Mexican health system com-

pared with those in other North American and Latin American countries. Changes in 2010 in the general health law (Cámara de Diputados del H. Congreso de la Unión, 2010) represent an attempt to increase health spending and improve efficiency. The same law and other reforms have also focused on expanding health-care and social security coverage, which, in Mexico, is not universal and is highly fragmented; a population of particular interest for the government is individuals in the informal sector, and some of these policies are trying to generate incentives for them to move from the informal to the formal sector.[2]

Whither U.S.-Mexican Relations?

In reviewing international and national policy set by Mexican and U.S. decisionmakers for the previous two sections, it became clear that the perspective of U.S. citizens is important because the United States is the top recipient for legal and illegal Mexican immigrants, and citizens set the tone for prospective new policies. Our review of literature and surveys of U.S. public opinion on immigration and trade reframed some important themes. The findings suggested that, although U.S. citizens recognize that there are solid achievements on which to build, some areas remain in need of further action.

U.S. Opinion on Immigration and Immigrants Is Mixed

Opinion polls show that a majority of Americans think that immigration is good for the United States and that immigrants contribute to the country and work as hard as or harder than U.S. natives. On the other hand, when asked specifically about illegal immigration, most respondents said they would like to see it reduced, and more than two-thirds think that it weakens the economy. The population is divided regarding how to handle illegal immigrants already in the country; slightly more than half of Americans think they should be required to

---

[2] We define *informal sector* as encompassing the economic activity of wage earners and self-employed individuals who do not make social security contributions (Aguila et al., 2011).

go home, while 40 percent said they should be allowed to stay in the United States (Pew Research Center for the People and the Press and Pew Hispanic Center, 2006).

Immigration Policies in Both Countries Continue to Be Characterized by Unilateralism

The United States focuses its efforts on securing its borders and limiting the expansion of the undocumented-immigrant population; Mexico, on the other hand, has focused on creating linkages with Mexican migrants, in order to improve their well-being and establish cooperation between them and their communities of origin.

Illegal Immigration Is a Charged Subject for U.S. Citizens and Mexican Immigrants Alike

Because illegal immigration has become an important topic in the U.S. public debate, several proposals for changes to immigration policies have been suggested. There are politically active groups with different interests in this debate, so immigration proposals usually include a combination of different policies to accommodate their concerns. In addition to increased border enforcement, which is an essential component of post–September 11 immigration bills, three major policy options have been debated: guest worker program, earned legalization, and legalization. All attempts for immigration reform since 1986 have failed to gain the support of enough legislators to pass, so the Immigration Reform and Control Act (IRCA) (Pub. L. 99-603, 1986) continues to be the basis of U.S. policy on illegal immigration.

Not All U.S. Citizens Believe That NAFTA Works in Their Favor

Although Americans seem to subscribe to the overall benefits of free trade as represented by NAFTA, different opinion polls run from 2004 to 2008 show that about 50 percent of respondents support renegotiation of NAFTA and believe that free-trade agreements take jobs away from Americans.

Resolution of Troubled Trucking Legislation Demonstrates That the Two Countries Can Overcome Contentious Bilateral Issues

NAFTA initiatives, designed to remove barriers to trade between the United States, Mexico, and Canada, called for the opening of the U.S. border to free-flowing truck and bus traffic from and to both neighboring countries from 2000 onward. However, citing highway safety and environmental concerns, the United States limited Mexican trucks in operation to commercial zones along the border. This is one factor that hindered the production-sharing nature of NAFTA by concentrating industry in the six states along the U.S. border and Mexico City. Moreover, retaliatory tariffs imposed by Mexico on U.S. imports further curtailed free trade. However, in July 2011, after extensive negotiations, an agreement was signed by the Mexican and U.S. governments that allows trucking and bus traffic for a trial period of three years (U.S. Department of Transportation, 2011).

Conclusions and Policy Implications

Taken together, our social and economic data provide insight into potential areas of concern for both Mexico and the United States and permit us to see important implications for policy.

Recommendations Addressing Mexican Immigration to the United States

Improve the Facilitation of the Legal Labor Market. Building a strictly legal labor market will be key to enhance the United States and Mexico's future relationship in the coming years. Mexico has a strong base of human capital that frequently seeks out a broader base of economic activity through employment in the United States. Unfortunately, many immigrants enter illegally and are hired illegally, forming an underground or informal labor market within the United States. Also, the system of issuing U.S. visas should be restructured and streamlined to meet the labor needs of legal industries, especially in the case of those industries seeking the help of low-skilled Mexican laborers.

Use Information Tools to Understand Migration Flows and Trends. U.S. and Mexican officials should work together to better understand population flows and trends. Databases that record individuals' unique labor histories in the United States and Mexico, along with information pertaining to qualifications, employment and unemployment periods, family characteristics, and contributions to U.S. and Mexican social security systems, might be improved to help understand migration flows.[3]

Have a Single Organization Committed to Recording Labor Movements of Immigrants to Support All Current and Future Immigration Policy. To the extent that the United States and Mexico might work more closely together to decrease illegal border activity, a correlating organization might specifically regulate and monitor all labor movements of individuals between Mexico and the United States. Although a binational organization might seem a rather futuristic ideal for U.S.-Mexican relations, it nonetheless has the potential of fulfilling long-term needs of both nations. A binational organization may be an effective solution, requiring strong commitment, collaboration, and resources from both countries.

Improve International Understanding When Approaching the Immigration Issue. Many U.S. policies address immigration without an understanding of the economic and social origins and dynamics of the phenomenon. Additionally, there is little consideration of the conditions in the United States that promote constant influxes of immigrants. It is important that policymakers from both countries begin to understand the causes and consequences of Mexican migration and take into consideration the particular regional disparities and economic and social conditions behind citizens' decisions to migrate. Greater understanding will help facilitate some important policies and programs—for example, one that promotes employment in underdeveloped areas will help with a goal of retaining citizens in their places of origin. A proportion of the population will still migrate from Mexico to

---

[3] It is important to note that, in making this recommendation, our intent is that such data would be used only for research purposes and would not be used to identify or track specific individuals.

the United States because of networks or cultural traditions engendering the desire to move north. However, a thorough analysis aimed at understanding which types of migrants are likely to stay in the place of origin will help policymakers design appropriate economic incentives.

Recommendations Addressing Mexican Reform

Stimulate Growth in the Formal Economy of Mexico. Working conditions would improve for many more citizens if workers were able to move from the informal to formal economy because there are better retirement plans and medical coverage for workers in those sectors. Policies to retain individuals in the formal sector are hard to implement, but programs to generate employment and economic growth and improve a country's comparative advantages are some of the key elements to improve conditions in the labor market.

Broaden the Tax Base in Order to Improve Government Revenues and the Capability to Target Social and Economic Issues. Tax collection can be improved by providing incentives to states for collecting taxes and for firms to move to the formal sector.

Promote Greater Market Competition. The Federal Competition Commission (Comisión Federal de la Competencia México, or CFC) should be vested with real powers to prevent predatory pricing, divest monopolies, and implement leniency programs (World Bank, 2006a). So far, the CFC has very limited power to prevent and take action over monopolistic practices.

Invest in Expanding Refining Capacity and Natural-Gas Production. Most of the production and exploration of energy from PEMEX, the government monopoly, is under severe pressure to update and improve efficiency. Allowing foreign participation can provide immediate solution to these issues and improve government revenue.

Allow Private Producers into Some Segments of Mexico's Energy Sector. In 1992, the Mexican Congress amended the constitution to attract private investment and allow limited private generation of electric power, when the need to provide adequate electrical power to sustain growing electricity demand became clear. The telecommunication sector was successfully privatized in 1990 (Aspe Armella, 1993). Privatization might be one option of reform for the energy sector.

Effectively Restructure Labor Regulation to Meet Current Workforce Needs. The labor regulation in Mexico is out of date, and overly stringent regulations contribute to the large informal sector. Lack of flexibility for part-time jobs and reentering the formal labor market after retirement are some of the issues that affect the most vulnerable segment of the population.

Allow for Less Costly Hiring and Firing Practices by Firms in Mexico. By doing so, the process of innovation that depends on quick formation and dissolution of firms will be aided while still adhering to rules of due process.

Promote Economic Growth in Rural Areas. The OECD (2007b) finds that it is crucial to have more coordination among the institutions in Mexico responsible for rural policies. A better institutional arrangement would allow institutions in charge of rural development to exploit synergies and identify potential complementarities in their policies.

Make Education Quality a Primary Policy Objective as Soon as Possible. Improving educational quality should be a major policy objective that will require improvement in the larger infrastructure, as well as in the area of teacher preparation. Another challenge is to increase enrollment and retention rates for upper-level institutions. Any new policies addressing these issues must take into account the existing achievement disparities across indigenous, public, and private schools.

Coordinate Social Policy Across All Sectors. Only a few social programs in Mexico have had positive impact in reducing poverty, and the specific objectives of different programs overlap. An effective and efficient social policy would require the elimination of programs with overlapping objectives and concentrate efforts on those with the highest impact.

Reform Pension Systems with Cash-Flow Deficits. Pension systems represent an important government liability. The Mexican Social Security Institute (Instituto Mexicano del Seguro Social, or IMSS) was reformed in 1997, and the State Workers Security and Social Services Institute (Instituto de Seguridad y Servicios Sociales de los Trabajadores del Estado, or ISSSTE) for government employees was reformed in 2007 (Aguila et al., 2011). Other public-sector institutions, such as the oil and electricity companies, universities, and many local governments, need to give urgent attention to reform their pension systems and guarantee their future financial feasibility. Reducing the government burden from the pension systems could allow allocating more resources to policies that promote economic growth.

Recommendations Based on U.S. Public Opinion and Analysis of NAFTA

Improve International Understanding Through Media and Educational Forums. Improvement of U.S.-Mexican relations begins with a strong foundation of public understanding on both sides of the border. Presenting information on Mexican and American culture, society, and politics via meaningful media outlets and formal and informal instructional opportunities in both countries will strengthen this critical relationship by educating citizens about past, present, and future international policies and concerns.

Explore Geographical Expansion Opportunities to Better Facilitate the Maquiladora Program.[4] Much of the FDI flowing into Mexico has been concentrated in the six states along the U.S. border and Mexico City. In order to foster more-even development, Mexican policymakers should encourage the setting up of enterprises in other locations to take advantage of local comparative advantages. Continuing the effort of building a sound transportation infrastructure will mitigate the increased cost of being located away from the border.

---

[4] According to Brauer (undated),

> A maquiladora company is a Mexican corporation operating under a special customs treatment, whereby it may temporarily import into Mexico machinery, equipment, replacement parts, raw materials, and, in general, everything needed to carry out its production activities.

Maquiladoras were originally linked to U.S.-based plants and generally were foreign invested. Well after creating the maquiladora program, Mexico, in 1990, created a similar program for domestic producers, called the Program for Temporary Imports to Promote Exports (PITEX) (Cañas and Gilmer, 2007). In 2007, the two programs were merged under the Decree for the Promotion of the Manufacturing Industry, Maquiladora, and Exportation Services (Decreto para el Fomento de la Industria Manufacturera, Maquiladora y de Servicios de Exportacion, or IMMEX program) (van't Hek, Becka, and Mejia, undated). Plants in the program are often still referred to as *maquiladoras.*

Looking to the Future

Whether relations between the two countries improve or deteriorate in the future will depend on the policies adopted by the respective leaders. Common interests deserving of attention include trade, immigration, security, and investment. Ultimately, the future of relations between the two countries depends critically on approaches to policies adopted by the administrations of the two countries, taking advantage of new opportunities without the baggage of past missteps and suspicions.

Acknowledgments

We wish to thank Nelly Aguilera, José Almaguer, Eduardo Andere, Pedro Aspe, Javier Beristain, Herminio Blanco, Brent Bradley, Gustavo Carvajal, Adriana Caudillo, Rodolfo Corona, Pedro Flores Crespo, Vicente Fox, Larry Harrington, Tom Johnston, Lynn A. Karoly, Isaac Katz, Katherine Krumme, Luz Lajous, Dan Lund, David Madero, Roberto Madrazo, Miguel Mancera, Gabriel Martínez, Carlos Moreno, Ricardo Mújica, Lisa Quigley, Ernesto Revilla, José María Rivera, Scott S. Robinson, Eduardo Robledo, Mario Henry Rodríguez, Tonatiuh Rodríguez, Oscar Jaime Roldán, Luis Rubalcava, Marta Sahagún de Fox, James P. Smith, Graciela Teruel, James Thomson, Rodolfo de la Torre, Hector Vanegas, Jesús Velasco, Jaime Zabludovsky, and René Zenteno for their contributions and assistance to this study. We thank Gordon Hanson, Lynn Karoly, Linda Martin, Robert Pastor, Marc Rosenblum, and Joanne Yoong for improving this monograph with their detailed comments; Kate Giglio, who provided assistance in putting the final document together; and Mariana Garci-Crespo, Norely Martínez, Nelly Mejia, Ashley Pierson, and Yvonne Torres for excellent research assistance. We thank Stacie McKee for shepherding the book through production, Lisa Bernard for editing and formatting, Sandy Petitjean for her excellent artwork, and Eileen La Russo for the design. We are also grateful to C. Richard Neu for his support and guidance while overseeing this entire project. Any errors of fact or interpretation are the responsibility of the authors. We acknowledge the support of RAND's International Programs, especially its director, Robin Meili, for supporting the publication of this monograph.

Abbreviations

BANXICO	Banco de México, or Mexican Central Bank
bpd	barrel per day
CBP	Bureau of Customs and Border Protection
CDI	Comisión Nacional Para el Desarrollo de los Pueblos Indígenas, or National Commission for the Development of Indigenous Peoples
CEESP	Centro de Estudios Económicos del Sector Privado, or Private Sector's Center for Economic Studies
CETES	Certificados de la Tesorería de la Federación, or Federal Treasury Certificates
CETU	Contribución Empresarial de Tasa Única, or Flat Rate Business Contribution
CFC	Comisión Federal de la Competencia, or Federal Competition Commission
CLS	continuous linked settlement
COFEMER	Comisión Federal de Mejora Regulatoria, or Federal Regulatory Improvement Commission
CONAPO	Consejo Nacional de Población, or National Population Council

CONASAMI	Comisión Nacional de Salarios Mínimos, or National Commission of Minimum Wages
CONEVAL	Consejo Nacional de Evaluación de la Política de Desarrollo Social, or National Council for Evaluation of Social Development Policy
CPS	Current Population Survey
DHS	U.S. Department of Homeland Security
DoD	U.S. Department of Defense
DREAM	Development, Relief, and Education for Alien Minors
EAP	economically active population
EIA	Energy Information Administration
EIU	Economist Intelligence Unit
EMIF	Encuesta sobre Migración en la Frontera Norte de México, or Survey of Migration in Mexico's Northern Border
ERS	Economic Research Service
EU	European Union
FDI	foreign direct investment
FTA	free-trade agreement
FUNSALUD	Fundación Mexicana para la Salud, or Mexican Health Institute
GCI	*Global Competitiveness Index*
GDP	gross domestic product
IADB	Inter-American Development Bank

ICE	Bureau of Immigration and Customs Enforcement
IFE	Instituto Federal Electoral, or Federal Electoral Institute
IMCO	Instituto Mexicano para la Competitividad, or Mexican Institute for Competitiveness
IMF	International Monetary Fund
IMMEX	Decreto para el Fomento de la Industria Manufacturera, Maquiladora y de Servicios de Exportacion, or Decree for the Promotion of the Manufacturing Industry, Maquiladora, and Exportation Services
IMSS	Instituto Mexicano del Seguro Social, or Mexican Social Security Institute
INA	Immigration and Nationality Act
INEE	Instituto Nacional para la Evaluación de la Educación, or National Institute for Education Evaluation
INEGI	Instituto Nacional de Estadística y Geografía, or National Institute for Statistics and Geography
INS	Immigration and Naturalization Service
IRCA	Immigration Reform and Control Act
ISSSTE	Instituto de Seguridad y Servicios Sociales de los Trabajadores del Estado, or State Workers Security and Social Services Institute
KMZ	Ku-Maloob-Zaap
LFCE	Ley Federal de Competencia Económica, or Federal Law of Economic Competition

NAFTA	North American Free Trade Agreement
OECD	Organisation for Economic Co-Operation and Development
PAHO	Pan American Health Organization
PAN	Partido Acción Nacional, or National Action Party
PAYG	pay as you go
PEMEX	Petróleos Mexicanos, or Mexican Petroleum
PISA	Programme for International Student Assessment
PITEX	Program for Temporary Imports to Promote Exports
PPP	purchasing power parity
PRA	personal retirement account
PRD	Partido de la Revolución Democrática, or Party of the Democratic Revolution
PRI	Partido Revolucionario Institucional, or Institutional Revolutionary Party
R&D	research and development
RIA	regulatory impact assessment
SEDESOL	Secretaría de Desarrollo Social, or Ministry of Social Development
SEP	Secretaría de Educación Pública, or Ministry of Public Education
SHCP	Secretaría de Hacienda y Crédito Público, or Ministry of Finance and Public Credit

SINAIS	Sistema Nacional de Información en Salud, or National Health Information System
SSA	Social Security Administration
UNPD	United Nations Population Division
USA PATRIOT	Uniting and Strengthening America by Providing Appropriate Tools Required to Intercept and Obstruct Terrorism
USCIS	U.S. Citizenship and Immigration Services
VAT	value-added tax
WDI	World Development Indicators
WHO	World Health Organization
WTO	World Trade Organization

CHAPTER ONE

Introduction

U.S.-Mexican relations tumbled from optimism in the 1990s, spurred by the implementation of the North American Free Trade Agreement (NAFTA), to a turning away in the years immediately after the 9/11 terrorist attacks. However, starting in the second half of the 2000s, there has been renewed interest in the importance of U.S.-Mexican relations on both sides of the border. Public attention has been stimulated by cover stories in national media regarding the construction of a border fence, authorized by the U.S. Congress through the Secure Fence Act of 2006 (Pub. L. 109-367), to alleviate the influx of illegal immigrants, also known as *undocumented* or *unauthorized* immigrants. The provisions of NAFTA have come under increasing scrutiny, especially during the campaign leading up to the 2008 U.S. presidential election, wherein some candidates called for a renegotiation of this trade agreement. President Barack Obama promised to make immigration reform a top priority upon his election, and the alleviation of the violent activity affiliated with Mexican drug cartels has piqued the concern of President Felipe Calderón, members of U.S. Congress, the U.S. Department of Defense (DoD), and local border police.

The renewed interest in U.S.-Mexican relations comes at a time when fiscal restraints generated by a downturn in the U.S. economy make clear the contentious but dependent relationship between these two neighbors. The geographic proximity of the two countries makes political, economic, and social association inevitable, and a binational, cooperative approach is one way to support and strengthen this relationship.

Purpose of This Study

The primary intent of this monograph is to provide an objective, binational presentation of facts and analysis of issues that concern each country and to fully explore the economic and political ties that bind them. We define *binational* as a way of examining an issue that takes into account the concerns and interests of both Mexico and the United States. It is our belief that Mexican policymakers could benefit from a detailed discussion on the issues surrounding migration from Mexico, a primary U.S. concern. Migration to the United States is usually seen in Mexico as a matter of migrants looking for a better life for themselves and their families. However, the debate is more complex in the United States, where immigration is seen as an issue affecting national security, national identity, public finances, and the wages of low-skilled Americans. Mexican proposals for bilateral migratory negotiations should acknowledge these topics in order to improve their likelihood of being received well by their U.S. counterparts.

Similarly, U.S. policymakers might benefit from an analysis of the economic and social situation in Mexico. Although Mexico continues to be classified as a middle-income country, the continued rapid growth of Asian countries has posed a serious challenge to Mexico's place in the emerging world economic order. It is in the interest of the United States to see a rapidly growing and stable Mexico, not only because it reduces the pressure on Mexican migration but also because trade with and investment in Mexico are likely to be beneficial to the United States. Economics is not a zero-sum game; the United States and Mexico can seek win-win arrangements.

Given that both countries have presidential elections in 2012, the time is ripe for critical review of long-term issues that have not been addressed. How does immigration to the United States affect the Mexican and U.S. economies? What do the social conditions of Mexico have to do with immigration? How do U.S. citizens feel about the policies that are in place? In this monograph, we address these broad issues to provide insight into the overarching questions generating much policy concern: What are the long-term trends underlying the state of U.S.-Mexican relations, and what is the future?

Approach of This Study

We took a multidisciplinary approach in pursuing answers to these questions in ways that were appropriate to the binational goal of this project. We sifted through a vast amount of evidence that already exists on Mexican immigration, Mexican social and economic conditions, and U.S. public opinion regarding relations with Mexico. We also conducted interviews with key policymakers in Mexico, addressing macroeconomic and microeconomic conditions, competition regulation, trade, labor markets, social security systems, social policies, and poverty alleviation. These discussions provided deep insight into the economic and social situation of Mexico, government emigration policies, and future challenges. We also carried out a thorough analysis of the primary social and economic issues of Mexico through examination of literature from both sides of the border, as well as survey data, country reports, and government documents regarding programs and policies.

Our aim was to analyze secular trends of the past few decades that relate to structural issues rather than the current situation, which continues to be in flux worldwide at the time of this writing. Although we make note of available data from the past few years, we focus more on events before the 2007 onset of the economic recession. This will ensure that short-term cyclical effects of the downturn and its aftermath do not overshadow longer-term issues. Such quantitative and qualitative assessment provides an understanding of major U.S. and Mexican policy achievement and challenges and allows us to weave policy suggestions throughout the monograph.

Our work here is somewhat similar to two monographs that also seek to provide a comprehensive view of Mexico and the impact of its policies on the United States. In *Bordering the Future: The Impact of Mexico on the United States* (2006), John A. Adams discusses the trends in U.S.-Mexican relations and their implications for American business and policymaking. Adams covers the following themes, which he feels affect the United States: agriculture, maquiladoras, immigration, energy needs, and the political dynamics in Mexico. He also provides an analysis of how Mexico and China compete with each other for American investment, trade, and resources for economic development. In *Good Intentions, Bad Outcomes: Social Policy, Informality, and Economic Growth in Mexico* (2008), Santiago Levy jointly examines the phenomena of macroeconomic stability, slow growth in formal employment, and the expansion of social programs in Mexico. He argues that differences in financing social security for workers in the formal sector and social protection programs for those workers in the informal sector amount to a tax on the former group and a subsidy to the latter group. This increases the incentive to stay in the informal economy and reduces aggregate productivity while contributing to persistent poverty.

However, although these books examine a few issues in depth, we aim to be more comprehensive. For instance, we consider the causes of immigration, characteristics, origins and destinations of migrants, and how regional disparities affect migration. We also examine the effects of tax policy, fiscal federalism, and the state of health, education, and poverty, issues that are not the focus of the aforementioned works. Our emphasis is also more squarely on policy—both assessing the progress made and outlining the challenges that remain. The approach of anchoring the attention of Mexican policymakers on U.S. concerns about immigration, and that of U.S. policymakers on Mexico's concerns of stagnation, poverty, and inequality, also distinguishes our work from others.

Clearly, there are other topics that are of interest to the United States and Mexico, such as climate change and cultural ties. But the topics we have chosen—immigration from Mexico and Mexico's economic and social development—are comprehensive and complex enough to illustrate the progress made by both countries and the challenges that remain. We have not considered the third partner of NAFTA—Canada—due to limitations of time and resources. However, we realize that the formulation of a comprehensive policy approach to Mexico that addresses many of the chronic issues discussed in our study would also have to include Canada in the equation.[1]

---

[1] In the North American summit, held in Mexico on August 10, 2009, four of the leading issues discussed were immigration, trade, security, and climate change. We deal extensively

How This Monograph Is Organized

This monograph is organized into four parts. The first, "Migration from Mexico: A Critical American Issue," is an examination of the history and patterns of migration from Mexico into the United States. It discusses the places of origin of Mexican immigrants and their destinations in the United States, as well as their effect on the U.S. and Mexican economies. It analyzes the reasons for this migration—the "pull" of the U.S. economy and the "push" of limited economic opportunities in Mexico. It concludes with an examination of policies toward immigration in both countries. Although policymakers from both countries would benefit from the analysis presented in that part, it is especially intended for Mexican policymakers to see a pressing issue through the eyes of their neighbor.

The second part, "Progress and Challenges: Mexico's Economic and Social Policy," discusses economic and social reforms in Mexico and the challenges that remain on these fronts. Although such an analysis is likely to be helpful to Mexican policymakers, it is also intended to illustrate for U.S. policymakers Mexican frustration over low growth and persistent poverty and inequality. This part first presents an overview of the economic and social situation in Mexico. It then delves deeper into a few areas of greatest concern in economic policy: competitiveness, taxes, labor, energy, fiscal federalism, and financial stability. This is followed by a discussion of social policy, poverty, and inequality. Health, education, social security, and social programs are all considered.

The third part, "The Past and Present of U.S.-Mexican Relations," surveys U.S. public opinion and developments on the immigration, economic, and trade fronts to assess the state of relations between the two countries.

The fourth and final part includes our conclusions and recommendations. The monograph ends with an appendix describing political history in Mexico.

---

with the first two in this monograph. Security is dealt with in a companion piece to this monograph (Schaefer, Bahney, and Riley, 2009).

PART ONE

Migration from Mexico: A Critical American Issue

Natives of Mexico account for the largest number of U.S. immigrants since the 1980s (1960–2000 U.S. Census, accessed via Ruggles et al., 2010). Although immigration has been a continuing topic of discussion in both countries since the early 1900s, its specific aspects—such as the size of migration flows, characteristics of migrants, immigration's causes, consequences, and subsequent policies—have evolved considerably over time. In the United States, the topic of immigration has been a major focus of public debate for many years, before being overtaken by the economic crisis that started in 2007. Although there has been some attempt to address this issue from a balanced perspective, solutions addressing the myriad of security, economic, and social issues surrounding immigration have been scarce.

Part One of this monograph contributes to the immigration debate by surveying related literature and discussing some of the primary points necessary to formulate sound immigration policies.

The first part is broken into six chapters. In Chapter Two, we offer statistical and demographic estimates of past and future Mexican immigrants moving into the United States. In Chapter Three, we present a historical overview of U.S. and Mexican migration policies. Chapter Four provides information about the sending and receiving states in the United States and Mexico, as well as characteristics of Mexican immigrants. Chapter Five provides a summary of previous studies regarding the effects of Mexican immigrants in the United States, and Chapter Six examines the main causes of Mexican emigration. The first part concludes with a discussion of some of the main challenges and opportunities in migration between the United States and Mexico.

CHAPTER TWO

Immigration by the Numbers

How many Mexican immigrants were working in the United States in the 1990s and 2000s? How many can the United States expect in the upcoming decades? This chapter addresses questions surrounding present and future Mexican immigration by highlighting the role of population growth, as well as the demographic transition in Mexico.

Immigration Statistics

Immigration is a phenomenon that involves millions of people. The number of Mexican immigrants in the United States grew dramatically in the 1990s and continued growing into the 2000s, as is shown in Figure 2.1. The Mexican-born population in the United States was estimated at 11.5 million in 2009, up from just 4.3 million in 1990 (Figure 2.1). In 2005, almost 16 percent of Mexico's working-age population resided in the United States (Giorguli Saucedo, Gaspar Olvera, and Leite, 2006).

Two remarks are important regarding these estimates. First, estimates of the Mexican-born population in the United States vary across sources, and those shown in Figure 2.1 might be low.[1] Second,

[1] For example, the U.S. Census Bureau recorded 9,177,487 Mexican-born residents of the United States in 2000, compared with the Current Population Survey (CPS) count of 8,072,288 (U.S. Census Bureau, undated [b]). Furthermore, the Pew Hispanic Center estimates that, in 2004, the United States had 11.2 million Mexican-born residents, compared with 10.7 million Mexican-born residents recorded in the CPS (Passel, 2005). Figure 2.1 uses CPS figures estimated by the Migration Policy Institute (undated).

Figure 2.1
Total Number of Mexicans in the United States, 1960–2009

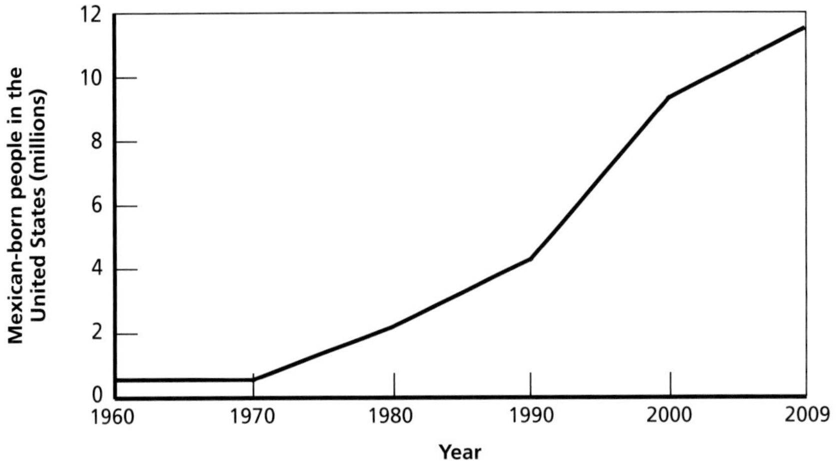

SOURCE: Migration Policy Institute, undated.
RAND *MG985/1-2.1*

these numbers do not differentiate between legal immigrants and illegal immigrants—the latter generally being defined as those who have entered the United States without following the appropriate U.S. rules or who entered legally but have overstayed their visas. Getting an accurate count is extremely difficult for two reasons: Illegal migrants have little incentive to declare their status, and they are difficult to pick up in large-scale national surveys that do not inquire about legality.

Nonetheless, there are a variety of estimates indicating that the number of illegal immigrants grew rapidly in the 1990s and first few years of the 2000s. In 1990, there were an estimated 3.5 million to 4.7 million illegal immigrants, of whom 1 million to 2 million were Mexican. By 2000, the total was estimated at between 7 million and 10.9 million, of whom 3.9 million to 4.8 million were Mexican (Hanson, 2006). Passel and Cohn (2009) estimate that, in 2008, 59 percent of illegal immigrants living in the United States came from Mexico. Although the common perception is that illegal immigration has increased constantly since the 1990s, estimates from the late 2000s indicate that the number of unauthorized immigrants has stabilized or even declined (Passel and Cohn, 2008b, 2010; Hoefer, Rytina, and Baker, 2009).

Many U.S. citizens of Mexican origin come from the pool of illegal entrants. From 1992 to 2002, about 56 percent of all legal Mexican immigrants were so-called *status adjusters*, people who had entered the United States illegally or who were in the United States on temporary visas (Hanson, 2006). Mexican immigrant flows have also spread widely from their traditional states of concentration (Arizona, California, Florida, Illinois, and Texas). The share of those born in Mexico living in the United States who are in these states declined considerably between 1994 and 2010, from 90 percent to just below 72 percent (Figure 2.2) (U.S. Census Bureau, 1994, 1995, 1996, 1997, 1998, 1999, 2000, 2001, 2002, 2003, 2004, 2005, 2006b, 2007b, 2008b, 2009c, 2010).

Figure 2.2
Share of Mexican-Born U.S. Population in Top Five Receiving States (Arizona, California, Florida, Illinois, and Texas), 1994–2010

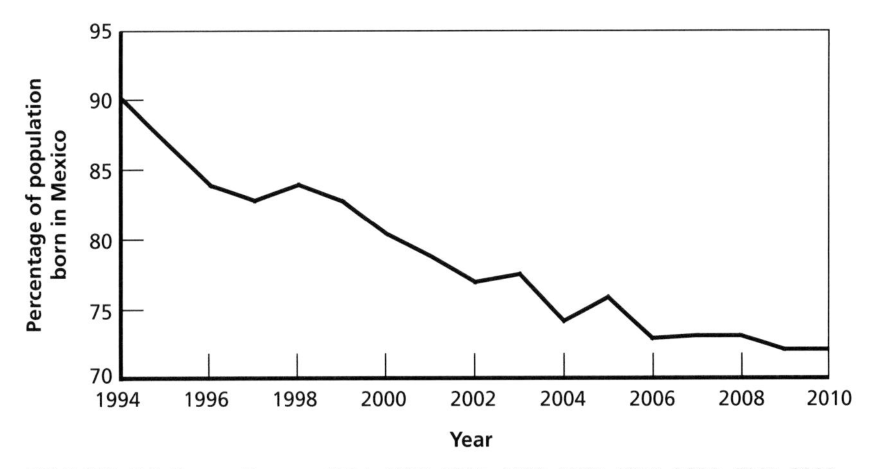

SOURCES: U.S. Census Bureau, 1994, 1995, 1996, 1997, 1998, 1999, 2000, 2001, 2002, 2003, 2004, 2005, 2006b, 2007b, 2008b, 2009c, 2010.

RAND *MG985/1-2.2*

Number of Future Mexican Immigrants

The number of future Mexican immigrants to the United States is dependent on a variety of factors, including future demographic, economic, and political developments. Focusing on demographic trends, some authors have forecasted that migratory flows—that is, the number of individuals coming anew into the country—might decrease. First, the large decline in birth rates driven by the Mexican government since the 1970s—fertility dropped from 6.5 children per woman to 2.2 between 1970 and 2005 (United Nations Population Division [UNPD], 2007)—might result in lower migration because the number of persons entering the labor force will drop significantly in the coming years because the largest cohort of the Mexican baby boom is entering the labor force around 2005 to 2015. Thereafter, the future cohorts are smaller (see, for example, Figure 8.8 in Chapter Eight). Second, it is argued that lower fertility implies smaller household size, which helps children stay in school longer, and educated Mexicans are less likely to migrate (P. Martin, 2002). However, this phenomenon is not expected to affect migratory flows before 2020 (Escobar Latapí and Martin, 2008). Third, although, in the past three decades, the Mexican labor supply grew at a much faster rate than the U.S. native-born labor supply, the respective growth rates will become much more similar in the following decades, easing the role of labor supply pressure in Mexico-to–United States migration (Hanson and McIntosh, 2009). It is important to note, however, that Mexico's working-age population as a share of the total Mexican population will continue to rise in the next two decades (Sedano, 2008), so demographic pressure on emigration will continue during this period. It is only in the decades following 2030 that the largest birth cohorts are reaching retirement age. However, even in 2050, the share of the working-age population is projected to be only 1 percentage point lower than it was in 2005.[2]

[2] The working-age population (15- to 64-year-olds) as a share of the entire population was 63.3 percent in 2005 and 64.9 percent in 2010 and is projected to be 66.7 percent in 2020, 66.0 percent in 2030, 64.1 percent in 2040, and 62.1 percent in 2050 (U.S. Census Bureau undated [b]). Therefore, even when the working-age population starts declining around 2030, the share of the working-age population by 2050 will be just 1 percentage point lower than it was in 2005.

Thus, provided that the percentage of working-age individuals who choose to emigrate stays constant, this would lead to an increase in absolute levels of Mexican immigration into the United States until 2030. However, other factors, such as economic growth, the development of both countries' labor markets in terms of demanded skills and offered wages, a changing skill composition of the Mexican population, and more-open or more-restrictive immigration policies, might also affect future immigration flows. For instance, evidence produced from the end of the 1990s to 2011 indicates that Mexican immigrants in the United States are more likely to have completed ten to 15 years of education (and less likely to have less than nine years of schooling or a college education) than nonemigrants in Mexico. In other words, the least educated and the most educated are less likely to migrate to the United States. This evidence and analysis by others (Chiquiar and Hanson, 2005) are consistent with somewhat positive selection of Mexican immigrants to the United States. Thus, the presence of a more educated population in Mexico might not necessarily imply lower migration rates because recent reforms in Mexican education target increasing educational attainment through the primary and secondary levels, those groups that are more likely to migrate. Although, for instance, the Mexican National Population Council (Consejo Nacional de Población [CONAPO], 2006) predicts a large migration (annual migration flows of 365,000 to 505,000 until 2030) under a variety of scenarios, actual future migration flows could deviate strongly from these projections. Future return migration flows are equally likely to be influenced by economic and political characteristics. The U.S. recession that started in 2007 and the global economic crisis that commenced in 2008 and affected the United States and Mexico, however, did not appear to increase return migration; studies (Instituto Nacional de Estadística y Geografía [INEGI] [National Institute for Statistics and Geography], 2009; Rendall, Brownell, and Kups, 2010) show low levels of return migration from 2007 to 2009.

CHAPTER THREE

Moving Out: Historical Background of Mexican Migration Policy

In this chapter, we describe essential moments in Mexican immigration to the United States, from both the U.S. and Mexican points of view. The relationship between the two countries has evolved primarily according to the uncertainties that accompany mass migratory movement and the responses of institutions and policymakers on both sides of the border.

Immigration Policies in the United States

Throughout the 20th and 21st centuries, the United States has had changing attitudes and policies toward immigration. At different times, U.S. immigration policies have been driven by the demand for cheap labor by industrialists and farmers, by the concerns of its citizens who worry about immigration as a threat to American well-being, and, at yet other times, by both of these contradictory pressures. Furthermore, prior to 2011, the United States found itself in the uncomfortable position of seeking increased economic integration while insisting on separation.[1]

Early U.S.-Mexican Relations: A Period of Shifting Ideals

The United States has formally welcomed Mexican workers twice during its modern history. The first was a program allowing contract

[1] This section draws on Massey, Durand, and Malone (2002) and DeLaet (2000).

workers to enter between 1917 and 1921 in response to World War I, exempting them from the literacy tests that applied to most immigrants from other regions. The second, again in response to war, was the so-called Bracero Program, a binational treaty for the temporary employment of Mexican farmworkers in the United States. Originally envisioned as a temporary, World War II measure, this program was continuously extended due to the increasing demand for agricultural workers and ended up running from 1942 to 1964, allowing for the temporary migration of nearly 5 million Mexicans.

In 1965, the U.S. government made its first effort to limit the number of Latin American immigrants by amending the Immigration and Nationality Act (INA) (Pub. L. 89-236, 1965). The amendments created a new visa-allocation system for the Eastern Hemisphere and established an overall yearly quota in the number of visas for Latin America, the Caribbean, and Canada, forcing Mexicans to compete for a limited number of visas with other immigrants from the Western Hemisphere. Further amendments to the INA in 1976, 1978, and 1980 created even more restrictions for Mexican immigrants, preventing young U.S.-born children from sponsoring their parents' immigration, extending a 20,000-per-country immigrant quota to the Western Hemisphere, and creating a single worldwide cap in the number of available visas. Thus, between 1968 and 1980, the number of visas available to Mexicans fell from an unrestricted supply to only 20,000 per year. Moreover, not even the 20,000 visas were guaranteed due to the fixed worldwide quota, so Mexicans were actually competing against immigrants from all other countries.[2]

Controlling Illegal Immigration After the 1986 Immigration Reform and Control Act

The most important migration law change since the INA came in 1986, with the Immigration Reform and Control Act (IRCA)

---

[2] Several factors caused total legal immigration to be higher than 20,000 over those years. Mainly, the 20,000 limit did not apply to immediate relatives, and a 1977 court order gave Mexicans 144,000 visas that were originally destined for Cubans; these visas were given between 1977 and 1981. Although these two factors increased the number of Mexican immigrants, the 20,000 limit was still legally in place.

(Pub. L. 99-603), which contained both restrictive and expansive provisions that tried to address the concerns of most stakeholders in the issue of illegal immigration. Among several measures, IRCA

- authorized a 50-percent increase in the immigration enforcement budget
- imposed sanctions against employers who consciously hired illegal immigrants
- increased the budget to carry out work-site inspections
- authorized an amnesty for long-term illegal immigrants
- incorporated a special program to legalize undocumented agricultural workers.

Although IRCA's goal was to reduce illegal immigration, the amnesty and agricultural-worker provisions were legislative compromises needed to gain the political support necessary for its passage. Moreover, the impact of employer sanctions was weakened by widespread document and identity fraud, lack of work-site enforcement, the low likelihood of being sanctioned, and low penalties (Rosenblum, 2005). Some literature found that neither this reform nor increased efforts in border enforcement until 2005 had much, if any, effect on the migration behavior of illegal Mexican immigrants (Donato, Durand, and Massey, 1992; Cornelius and Salehyan, 2007).

As it became clear that IRCA was not slowing legal and illegal immigration as much as initially hoped, new revisions to immigration and domestic laws were passed in the 1990s.

The Immigration Act (Pub. L. 101-649, 1990) continued to rule most legal immigration as of 2011. It included provisions that increased the global cap in the number of visas given each year, expanded the number of visas given to skilled workers, and created a new category of "diversity" visas for individuals from countries underrepresented in post-1965 immigration—and, therefore, not for Mexicans. The Illegal Immigration Reform and Immigrant Responsibility Act (Pub. L. 104-208, 1996) increased penalties for illegal entry and alien smuggling, authorized wiretaps and undercover operations against alien smuggling, and provided funding to allow construction of fencing in San

Diego, the acquisition of new military technology, and the hiring of additional U.S. Border Patrol agents. And the Personal Responsibility and Work Opportunity Reconciliation Act (Pub. L. 104-193, 1996) contained provisions excluding illegal immigrants—with limited exceptions—from most federal, state, and local public benefits; required the Immigration and Naturalization Service (INS; since 2003, U.S. Citizenship and Immigration Services, or USCIS) to verify the legal status of immigrants before they could collect any federal benefit; and placed new restrictions on *legal* immigrants' access to public services.

Beyond 9/11

The terrorist attacks of September 11, 2001, led to changes in the law aimed at preventing and combating terrorism, with multiple implications for immigration policy. The Uniting and Strengthening America by Providing Appropriate Tools Required to Intercept and Obstruct Terrorism (USA PATRIOT) Act (Pub. L. 107-56, 2001) broadened the terrorism-related definitions in the INA, expanded grounds of inadmissibility to include immigrants who publicly endorse terrorist activities, and expanded the government's powers to monitor students in addition to legal immigrants and to detain and expedite the deportation of noncitizens suspected of links to terrorists. The Homeland Security Act (Pub. L. 107-296, 2002) established the U.S. Department of Homeland Security (DHS) and drastically restructured immigration services and enforcement agencies, including the INS and the Border Patrol, incorporating them into DHS (LeMay, 2004; Waslin, 2003). Immigration enforcement is now a function of the Bureau of Customs and Border Protection (CBP) and the Bureau of Immigration and Customs Enforcement (ICE), while USCIS is in charge of immigration service functions (Meissner and Kerwin, 2009).

Emigration Policies in Mexico

Like the United States, Mexico has had a contradictory attitude toward outward migration, which has resulted in different policies and opinions held by federal, state, and local governments. This lack of a unified policy stemmed from diverging views regarding the benefits and costs of emigration and from different policy goals and priorities at each government level. Furthermore, Mexican emigration policies have been severely restricted by the policies of the United States. Since the 1990s, Mexico has focused significant efforts on ensuring the good treatment of its nationals in the United States.[3]

Early Emigration Policy and Questions of Labor

During the first part of the 20th century, the Mexican federal government actively worked to stop emigration to the United States, perceiving it as a threat to national unity and a symbol of Mexico's weaknesses compared with the United States. At the time, U.S. law formally prohibited the entry of workers already holding labor contracts; in practice, there was nearly free immigration (Fitzgerald, 2006; Orth, 1907). The Mexican Constitution of 1857 established freedom of exit and travel, subject to some restrictions in criminal and civil matters, but state governors wanted to increase barriers to migration, so the 1917 Constitution included provisions specifying that municipal officials had to ensure that emigrating workers had signed contracts detailing wages, hours, and repatriation costs to be paid by the employer.

Soon after, worried about labor shortages, the border states of Sonora and Chihuahua prohibited the exit of workers. Tamaulipas and Jalisco also instituted restrictive measures following the new constitution. Further limitations were instituted in 1926 with a law allowing the federal government to prohibit migration of workers whose contracts had not been approved by the municipal president of their place of origin.

The differences between the desires of the national government and those of the local elites became apparent during the Cristero War (1926–1929) and follow-on events in the 1930s. In conflict areas, elites and local governments encouraged migration as a safety valve to prevent landless peasants from causing unrest. National policy then went through a significant reversal in the 1930s and 1940s. In the 1930s, the

---

[3] This section draws on Fitzgerald (2006).

government encouraged repatriation from the United States during the Great Depression.

The Labor Law and Legacy of the Bracero Program

In the 1940s, however, the Mexican government made a series of agreements with the United States facilitating temporary agricultural labor in what eventually became the Bracero Program. However, in the late 1940s through the mid-1950s, the Mexican government tried at times to halt all emigration due to disagreements with the U.S. government regarding contracts, wages, and working conditions. Mexico had little leverage to force the U.S. government to compromise, however, and the Bracero Program continued without major changes until 1964.

The Mexican government attempted to extend the Bracero Program after it ended and through the early 1970s. Eventually, Mexican policymakers grew comfortable with the large flows of undocumented migration that resulted from the 1965 U.S. changes to the INA, seeing them as an instrument to relieve the demographic pressure from Mexico's rapid population growth. In 1974, Mexico eliminated the penalties to emigrating workers who did not have a contract and notified the U.S. government of its decision not to seek a new Bracero Program, at a time when the United States appeared interested in establishing a new migratory agreement in exchange for privileged access to Mexican oil (Rosenblum, 2007; Fitzgerald, 2006). This increased acceptance of the migratory phenomenon by Mexican authorities accelerated in the 1990s, when the Mexican government and consulates in the United States started paying more attention to emigrant communities.

Migration Measures in Mexico Since 2000

During his 2000 presidential campaign, Vicente Fox pledged to govern "for 120 million Mexicans," referring to 100 million in Mexico and 20 million in the United States (Rosenblum, 2004). After he was elected president, his administration made reaching an immigration agreement one of the cornerstones of its U.S. policy. The government proposed four migration measures and one additional related measure. The migration measures included eventual legalization of illegal immigrants in the United States, a new guest-worker program, U.S.-Mexican measures to end border violence, and a higher ceiling for immigrant visa limits for Mexicans. The related measure included deeper integration along the lines of the European Union, including stronger efforts by Mexico to curb illegal immigration and more assistance from the United States to promote development in Mexico (Price, 2000; Castañeda, 2001; Fox, 2001; P. Martin, 2004).

Although achieving this ambitious bilateral migration agreement was an important focus of the Fox administration, this goal became unattainable after the terrorist attacks of 2001 shifted U.S. foreign policy toward national security (Fernández de Castro et al., 2007; S. Martin, 2003). After this setback, all negotiations stalled, and opportunities to pursue bilateral policies were effectively closed. A notable action of the Fox administration was the 2003 creation of the Instituto de los Mexicanos en el Exterior (Institute of Mexicans Abroad), which has as its main objectives to reach out to Mexican nationals living in other countries in order to improve their well-being and to establish cooperation between them and their communities of origin. This institute was preceded by the Program for Communities of Mexicans Abroad, founded in 1990 (Figueroa-Aramoni, 1999).

In the past two decades, some Mexican states have created government agencies focused on migrants and their sending communities. These agencies seek to establish contact with migrant organizations in other countries—mostly the United States—to provide them with assistance and involve them in policies to develop their communities of origin. Although some states, such as Jalisco, Michoacán, and Zacatecas, have been able to develop well-organized and active organizations, most others have not designed policies targeting the specific needs of their population and simply limit their activities to the implementation of federal programs (Fernández de Castro et al., 2007). Regardless, these state-level initiatives highlight a trend toward decentralization of migration-related policies, and some interesting results are starting to surface from the efforts of the most-organized states, such as Zacatecas, which now allows its migrants living abroad to run in state congressional elections and, in 2001, negotiated a program to manage the allocation of guest-worker visas (H2B) with the U.S. consulate in Monterrey (Fitzgerald, 2006).

CHAPTER FOUR

Immigration Patterns

In this chapter, we describe U.S. migration to Mexico, the primary destinations of Mexican migrants in the United States, and the sending regions in Mexico. Also presented are some of the primary social and economic characteristics of Mexican migrants.

Migration from the United States to Mexico

It is worth noting that there is a flow of migrants from the United States to Mexico. According to Mexican statistical authorities, a total of 343,591 Americans lived in Mexico in 2000, which corresponds to 79.7 percent of all foreign-born residents of Mexico and 0.35 percent of the total Mexican population (INEGI, 2007a). However, this figure might be low: The U.S. Department of State reported that more than 500,000 U.S. citizens were living in Mexico in 2007 (U.S. Department of State, 2010b). The latest figures available (as of this writing), from the Migration Policy Institute (2006, p. 24), estimate that the U.S.-born population in Mexico increased 84 percent between 1990 and 2000.

Migration from Mexico to the United States is mainly work-related; in migration from the United States to Mexico, retirement plays an important role. Although the vast majority of U.S. citizens living in Mexico are children with at least one Mexican-born parent, there is a significant trend of U.S. seniors moving to Latin America for retirement. The U.S. senior population living in Mexico increased 17 percent between 1990 and 2000, but some municipalities with retiree communities have experienced a much faster growth of this population; some notable examples are Chapala (581 percent), Los Cabos (308 percent), and San Miguel Allende (48 percent) (Migration Policy Institute, 2006).

Primary Mexican Migrant Destinations in the United States

Between 1994 and 2010, the number of Mexican-born residents of the United States grew by about 83 percent (Table 4.1). The percentage of the U.S. population that is Mexican-born residents grew from 2.5 percent to 3.9 percent as the total U.S. population increased from 260 million to 304 million. Although the inflow of Mexican immigrants to the United States slowed down during the economic downturn period of 2001–2003, it regained strength between 2004 and 2006, holding at relatively high levels and even increasing through mid-2006. There is some evidence of a slackening in migration since mid-2006 (Terrazas, 2010; Pew Hispanic Center, 2011).

As Table 4.1 shows, although the top five destination states of Mexican immigrants remained the same between 1994 and 2010, the Mexican-born population in three of the five top destination states actually grew more slowly than in the United States overall. In 1994, these five states held 88 percent of the Mexican-born population in the United States, but this figure had decreased to 72 percent by 2010. This is a clear indication that the Mexican-born population is spreading out across the United States.

California had 53.7 percent of total Mexican immigrants in 1994; by 2010, this figure had decreased to 39.9 percent. Texas retained a stable proportion of Mexican immigrants, 20 percent. Arizona increased in the proportion of Mexican immigrants from 4.1 percent in 1994 to 5.1 percent in 2010, with a 128-percent change in the number of Mexican-born residents; in Illinois, the proportion of Mexican immigrants increased by only 24 percent. There are strong differences among receiving states in the factors that affect migrant destinations, such as the composition of the Mexican sending states and

Table 4.1
Top Destination States for Mexican-Born Residents in the United States, 1994 and 2010

Destination	Total Mexican-Born Residents	Percentage Change, 1994–2010	Share of State Population	Share of Mexican-Born U.S. Population
1994				
California	3,480,232		10.9	53.7
Texas	1,296,080		7.1	20.0
Illinois	511,948		4.4	7.9
Arizona	265,436		6.6	4.1
Florida	195,712		1.4	3.0
United States	6,485,253		2.5	100.0
2010				
California	4,736,911	36.1	12.9	39.9
Texas	2,375,851	83.3	9.6	20.0
Illinois	634,606	24.0	5.0	5.3
Arizona	604,405	127.7	9.3	5.1
Florida	245,020	25.2	1.3	2.1
United States	11,872,688	83.1	3.9	100.0

SOURCES: U.S. Census Bureau, 1994, 2010.

existing social networks and demand for labor. These, along with other explanations, are discussed in more detail later in this monograph.

Migrant Origins in Mexico

Mexico's central-west region (Michoacán, Guanajuato, and Jalisco) has traditionally had the highest levels of out-migration (Bustamante et al., 1998). However, the importance of this region has somewhat diminished over time, as all Mexican regions began to participate in migratory flows, although to different extents. CONAPO reported that 96 percent of Mexican municipalities registered some level of migration in 2000, and 37 percent of them registered medium or high migratory intensity (Zúñiga Herrera, Leite, and Rosa Nava, 2004).

In addition to the three states mentioned above, Chihuahua, Guerrero, Durango, Mexico City, Tamaulipas, Zacatecas, and the state of México gained in importance as sources of migratory flows to the United States. As can be seen in Table 4.2, in the year 2000, these states accounted for half of Mexico's total population and three-fifths of total migrants. In the first decade of the 2000s, the south and southeast regions (including the states of Chiapas and Oaxaca) have become the most-important emigrant regions (Terrazas, 2010).

Distance to the United States is an important factor explaining the origins of migrant flows. As can be seen in Figure 4.1, except for the border states (marked with solid dots), there is a clear negative relationship between a state's distance to the United States and the percentage of households with migrants in that state.

Table 4.2
Population and Migrant Share of Top Ten Migrant-Sending States, 2000

Name	Share of Total Population in 2000 (%)	Share of Total Number of Migrants in 2000 (%)
Jalisco	6.5	10.9
Michoacán	4.3	10.5
Guanajuato	4.9	10.4
México	13.2	8.1
Guerrero	3.2	4.7
Zacatecas	1.5	4.2
Mexico City	8.8	3.8
Chihuahua	3.1	3.2
Durango	1.5	2.7
Tamaulipas	2.8	2.1
Total	49.7	60.5

SOURCE: 2000 Mexican population census data from INEGI.

Figure 4.1
Migrant Households, by Mexican State and Distance to the United States, in 2000

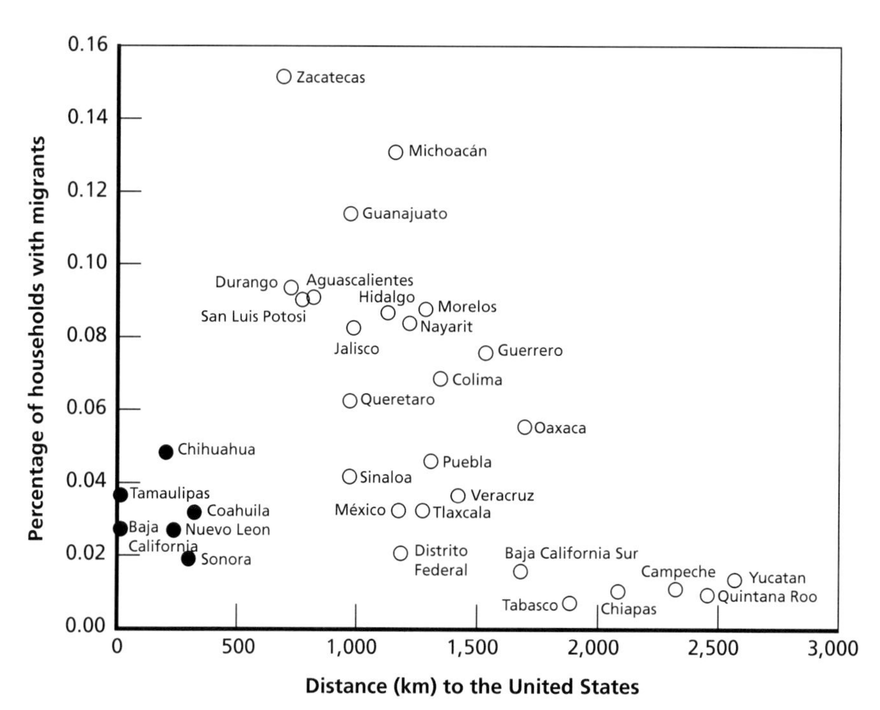

SOURCE: Hanson, 2006.
RAND *MG985/1-4.1*

Economic and Social Characteristics of Mexican Migrants

Despite the changing regionalization of Mexican migration to the United States, economic factors continue to be the main determinants of migration. The typical immigrant is most likely to be male and of working age. An estimated 70 percent of Mexican immigrants are between 15 and 44 years of age, and there are 24 percent more migrant men than migrant women. Mexican immigrants are significantly less educated not only than the U.S.-born but also than other immigrants in the United States: Eighty-six percent of Mexican immigrants have 12 years of schooling or less, while the same figure is 45 percent for both

U.S. natives and immigrants from other countries (Zúñiga Herrera, Leite, and Rosa Nava, 2004).

Given the schooling levels of Mexican immigrants, the common perception is that they come from the lowest socioeconomic levels in Mexico. However, there is evidence that migrants are, in fact, selected not from the lowest skill levels, with less than nine years of schooling, but rather from the middle of the skill distribution in Mexico, with ten or more years of schooling but no college degree. Table 4.3 shows the education levels of the Mexican resident population and three types of migrants identified by the Binational Study on Migration (1997). Mexican migrants, including individuals who intended to return to Mexico, had, on average, more years of education than the Mexican population, and migrants who settle permanently in the United States are considerably more likely to have at least a high school diploma. Although the distribution of educational achievements might have shifted considerably in the past 15 years among the different migrant and nonmigrant groups, this is consistent with the more-recent results in Chiquiar and Hanson (2005), who find that Mexican migrants are more educated than nonmigrants, falling in the middle and upper levels of Mexico's wage distribution.

Table 4.3
Education of Mexican Population and Mexican Migrants to the United States

Years of Education	Mexico Resident Population	Temporary Migrants with Permanent Residence in Mexico	Migrants (legal and illegal) Living Permanently in the United States	Naturalized Migrants (U.S. citizens)
Average years of education	5	6	8	—
Less than 5 (%)	46	39	28	24
5 to less than 12 (%)	44	52–60	48	43
12 or more (%)	10	1–9	24	33

SOURCE: Binational Study on Migration, 1997.

CHAPTER FIVE

Migration's Effects on Origin and Destination Countries

The impact of immigrants in the United States has been the main subject of the debate on immigration policy. A subject less discussed in the United States is the impact that migration has on families and the economy of sending countries, such as Mexico. In this chapter, we summarize the evidence on these two topics. In particular, we discuss the impact that migration has on wages in both countries, on the fiscal burden in the United States, and on household decisions, such as educational investment in Mexico. Other issues that are addressed are the impact that immigration has on economic growth and on demographic trends in the United States.

Immigration's Impact on the United States

Most Mexican immigrants in the United States work in low-skilled occupations. Among immigrants from all countries of origin, however, the fastest growth in employment occurred among middle-skilled occupations that require more than a high school diploma but less than a college degree (Capps, Fix, and Lin, 2010). Although unauthorized workers are employed in a variety of occupations, they are markedly underrepresented in white-collar occupations and overrepresented in services, farming, and construction (Passel, 2006; Capps, Fortuny, and Fix, 2007). Figure 5.1 shows that, overall, immigrants represent 15 percent of the total U.S. labor force and 45 percent of lower-skilled workers. Unauthorized workers represent 5 percent of the U.S. labor force and 23 percent of lower-skilled workers. In 2006, Mexican-born work-

Figure 5.1
Immigrant Shares of U.S. Population and Labor Force, 2005

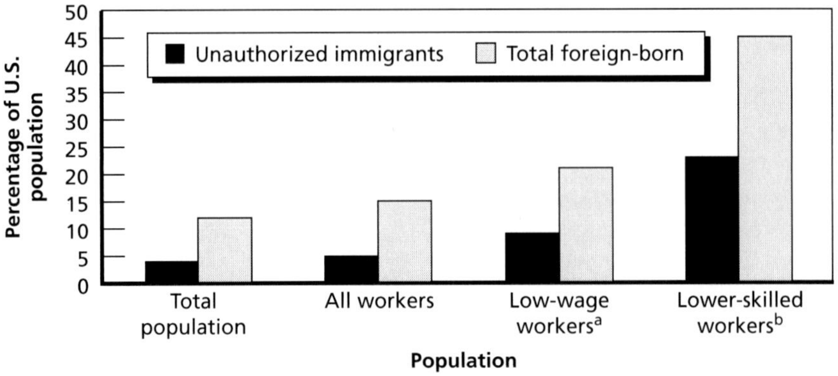

Percentage of U.S. population
50
45
40
35
30
25
20
15
10
5
0
Unauthorized immigrants
Total foreign-born
Total population
All workers
Low-wage workers[a]
Lower-skilled workers[b]
Population

SOURCE: Capps, Fortuny, and Fix, 2007.

[a] Workers who earned less than 200% of the minimum wage.
[b] Workers with less than a high school education.

RAND *MG985/1-5.1*

ers accounted for 31 percent of the immigrant workforce and 50 percent of immigrants who earned less than $20,000 per year (Gonzales, 2008). According to Capps, Fortuny, and Fix (2007), if the labor-market and immigration trends as observed in 2000–2005 continue, the U.S. low-wage workforce will become increasingly immigrant and somewhat more unauthorized.

The Presence of Immigrants Both Benefits and Challenges the Host Country

One of the most controversial aspects of immigration is its effect on the well-being of the native population. Much of the focus has been on immigration's effects on the wages of native workers with less than a high school education, with the expectation that the influx of immigrants with less education than the U.S. average would reduce the wages of those immigrants' native counterparts. Economists debate the best methodology for measuring immigration's effect on the wages of native workers, and different methods often produce different estimates.

However, there is an emerging consensus about the likely range of the size of the effect. For example, Smith and Edmonston (1997) summarize the available literature and find that the average effect is about a 4-percent decline in wages for native-born workers with less than a high school education. They state (p. 7), "immigration has only a relatively small adverse impact on the wage and employment opportunities of competing native groups."

A review by Friedberg and Hunt (1995, p. 42) reached a similar conclusion, noting that "a 10 percent increase in the fraction of immigrants in the population reduces native wages by at most 1 percent." Later work by Borjas (2003), using different statistical methods, estimated that, between 1980 and 2000, the wages of U.S. native workers without a high school degree, who represent about 8 percent of the native labor force (Hanson, 2009), fell by up to 9 percent due to immigration (Borjas, 2003). Although this estimate is among the highest produced, other studies have generated smaller estimates in line with earlier research (see, for example, Ottaviano and Peri, 2008).

Card (2009, p. 5) acknowledges the controversy surrounding the appropriate framework to measure the effect of immigration on native wages but still concludes that "immigration has had very small impacts on wage inequality among natives."

In addition, immigration of low-skilled workers might also benefit natives by lowering the prices of labor-intensive goods and services. For example, Cortes (2008) finds that a 10-percent increase in the share of low-skilled immigrants in the labor force decreases the price of immigrant-intensive services by 2 percent, due to the reduction in wages. Palivos (2009) also argues that illegal immigration has positive effects on the receiving country because illegal immigrants are paid less than the value of what they produce, leading domestic households to increase their holdings of capital.

The impact that immigration has on U.S. economic growth has been addressed relatively infrequently in empirical studies, and there is no consensus on the strength or even the direction of this impact. Peri (2009), using U.S. state-level data for each census year between 1960 and 2006, finds that immigration does not affect native employment and increases productivity per worker and that each 1-percent increase in employment due to immigration raises productivity per worker by 0.5 percent in that state. In combination, these two effects would increase per capita economic growth.

In addition to its economic impacts, immigration also has consequences for the demographic development of a receiving country. The most evident change is that, in a situation in which fertility rates of the native-born remain unaltered and the composition in terms of region of origin of new immigrants differs from that of previous immigration waves, the ethnic composition of the United States will be altered. Recent projections show that, by 2050 the white non-Hispanic population may have become a minority (Passel and Cohn, 2008a), and Coleman (2006) judges this change, and a similar one in Europe, to be sufficiently significant to label it the "third demographic transition." A second effect is the growing population size: It is estimated that the U.S. population will grow from 296 million to 438 million from 2005 to 2050 and that 82 percent of this increase, as well as the entirety of the increase in the working-age population, will be due to new immigrants arriving during this period and their descendants (Passel and Cohn, 2008b). There are also more-indirect effects. For instance, international immigration can induce internal migratory movements of native-born individuals. Empirical studies do not offer a clear picture of whether this is the case: Frey and Liaw (1998) and Borjas (2006) state that immigration is associated with higher internal out-migration and reduced native-born in-migration, while Kritz and Gurak (2001) and Card and DiNardo (2000) do not find evidence of internal out-migration.

The Immigrant Workforce Has Only a Minor Effect on Government Burden

Another issue debated about migrants' effects on their receiving countries is the balance between taxes they pay and the cost of government services they receive. Although conceptually this is a simple calculation (whether immigrants pay more in taxes than they receive in government benefits), it is difficult to estimate in practice. One important study on the total fiscal impact of immigrants is Smith and Edmonston (1997). They estimate the fiscal impact of all immigrants—legal and illegal—on federal and state budgets.[1] They find positive fiscal benefits across all levels of government for the average immigrant. Using data from Smith and Edmonston (1997), the Council of Economic Advisers (2007, p. 645), concludes, "Although subject to the uncertainties inherent to long-run projections, careful forward looking estimates of immigration's fiscal effects, accounting for all levels of government spending and tax revenue, suggest a modest positive influence on average."

In a summary of research in this area, Hanson (2007) suggests that the total fiscal impact of immigration is modest, although it is likely that much of the burden is borne by state governments. Fiscal impacts of immigration tend to be positive on federal budgets, at least partly because illegal and temporary immigrants are ineligible for most social programs and permanent residents need to be present in the country for five years before they become eligible (Assistant Secretary for Planning and Evaluation, U.S. Department of Health and Human Services, 2011). They tend to be negative on state and local budgets. Indeed, Smith and Edmonston (1997) note that the net taxpayer benefit declines with the education of immigrants and can become a burden in a state that, like California, has young, low-income immigrants who have children in schools. For instance, for California, they estimate that the average immigrant-headed household receives an annual net transfer from the state and local sectors of $3,463 (Smith and Edmonston, 1997, p. 280).

In summary, the available evidence suggests that, whether positive or negative, the total fiscal impact of immigration on the U.S. economy is modest (Hanson, 2007; Escobar Latapí and Martin, 2008; Card and Lewis, 2005; Council of Economic Advisers, 2007; Card, 2007; Meissner et al., 2006).

---

[1] The fiscal impact is an estimated present value over the expected lifetimes of immigrants with both legal and illegal status, including the expected lifetimes of their descendants. The computation includes all state and local taxes and all state and local programs, such as incarceration costs, Medicaid, and welfare programs.

Immigration's Impact on Mexico

In various ways, international migrants maintain ties to their countries of origin. Although there is evidence that early European migrants to the United States maintained similar connections, these ties were mostly ignored in records of immigrant history (Schiller, Basch, and Blanc, 1995). Scholarly literature on migration has increasingly focused on transnational migrants, those who keep attachments both to the country to which they have migrated and to their country of origin (Levitt, 2004). Much of this literature has focused on remittances and their potential use as development tools (see Chapter Ten in Part Two), but migration's impact on sending countries is broader, affecting social, cultural, and economic structures (Guarnizo, 2003; Itzigsohn and Giorguli Saucedo, 2002; Gould, 1994). For instance, although migration has increased household income, it has also altered family composition, causing parental absenteeism and a more important role for grandparents and other family members in the education of children (Hanson and Woodruff, 2003; McKenzie and Rapoport, 2006; Creighton, Park, and Teruel, 2009; Tanner, 2010).

Given Mexico's geographic proximity to the United States and the long tradition of circular migration between the two countries, it is not surprising that Mexican migrants maintain strong ties to their home communities.

Immigration Prospects Are Broadened

The first important effect of this continuous attachment is a resulting process of social network growth on both sides of the border. Massey and García España (1987) conducted an analysis of the development of social networks that arise from Mexican migration to the United States; they conclude that social networks reduce the cost of migration for other groups of nonmigrants, inducing them to migrate and thus perpetuating the process. In short, migration combined with sending-country ties generates further migration, and this process continues independently of the causes that originated the first migratory movements.

Social Networks Increase the Chances That the Poor Will Migrate

Social networks, and the resulting lower costs of migration, have other important effects. McKenzie and Rapoport (2007) find that Mexican migrants tend to come from the middle of the wealth distribution when social networks are not well developed. However, as community migrant networks grow, migration becomes less costly, and the poor become more likely to migrate. Because migrant households become richer due to income earned abroad, large migrant networks "spread the benefits of migration to members at the lower end of the consumption and wealth distributions . . . thereby reducing inequality" (p. 22). McKenzie and Rapoport (2007) also find evidence that migration benefits the upper-middle portion of the wealth distribution when migrant networks are small.

Similarly, Munshi (2003) finds that, for individuals from well-established migrant communities, the likelihood of a migrant being employed and holding a preferred nonagricultural job is higher when his or her migration network is larger. Moreover, Munshi finds, the most–economically disadvantaged members of the community, such as women, the elderly, and the least educated, benefit most from a larger network.

Wages Are Increased at Home as Skill-Group Members Leave

In addition to its direct income effects due to earnings abroad, migration has other effects on inequality due to externalities. Mishra (2007) finds that a 10-percent decrease in the number of Mexican workers in a skill group—due to migration—increases average wages in that skill group by about 4 percent. However, this impact that migration has on wages is not homogeneous across education groups, such that Mexicans with 12–15 years of education benefit the most from the migration of their peers. Mishra concludes that migration has contributed to the observed increase in wage inequality in Mexico since the mid-1980s. Similar results, in terms of both the size of the wage impact and the largest benefit for those in the middle of the skill distribution, are found by Aydemir and Borjas (2006).

Households Shift Priorities

Finally, migration might have unintended consequences on other aspects of households' decisionmaking. In particular, McKenzie and Rapoport (2006) examine whether the prospect of future migration for children who grow up in a household in which at least one member is an international migrant might actually lower the incentive to invest in education. They find evidence consistent with this effect: When living in migrant households, 12- to 18-year-old boys are 22 percent less likely to complete junior high school and 16- to 18-year-old girls are 13 percent less likely to complete high school. This is certainly a cause for concern because these negative effects appear to outweigh the positive impact that remittances have on educational attainment; in many cases, households have more income, which allows children to attend school instead of working. Cash-transfer programs (see Chapter Ten in Part Two) are a promising policy alternative to counteract this effect: Angelucci (2005) finds that school-age children in communities that receive transfers that are partially conditional on their continued school attendance are less likely to migrate than children in communities not participating in the program.

CHAPTER SIX

Causes of Migration from Mexico to the United States

We determined three categories for the main factors that cause migratory flows:

- *Demand-pull* includes recruitment by U.S. employers or significant job availability in the United States, in addition to significant wage differentials between the two countries.
- *Supply-push* includes poor performance of the Mexican economy and strong regional socioeconomic inequalities in Mexico.
- *Networks* include family members and friends who already live in the United States or other channels that facilitate Mexican migration (P. Martin, 2002).

Demand-Pull: U.S. Economic Factors

U.S. Economic Conditions Are Critical to Decisions to Immigrate

Job availability in the United States is generally considered the main determinant of short-term immigration flows (Escobar Latapí and Martin, 2008; Passel and Suro, 2005). Immigrant flows usually follow U.S. economic conditions, such as the sustained growth in the 1990s, and the economic cycles during the 2000s (see Figure 6.1).

Wage Differences Between the Two Countries Also Play a Role

In Figure 6.1, we observe a high correlation between U.S. gross domestic product (GDP) growth and the change in immigrant flows from Mexico, particularly since 2000. Demand-pull factors also include

Figure 6.1
Annual Change in Immigrant Flows to the United States and U.S. Gross Domestic Product, 1991–2010

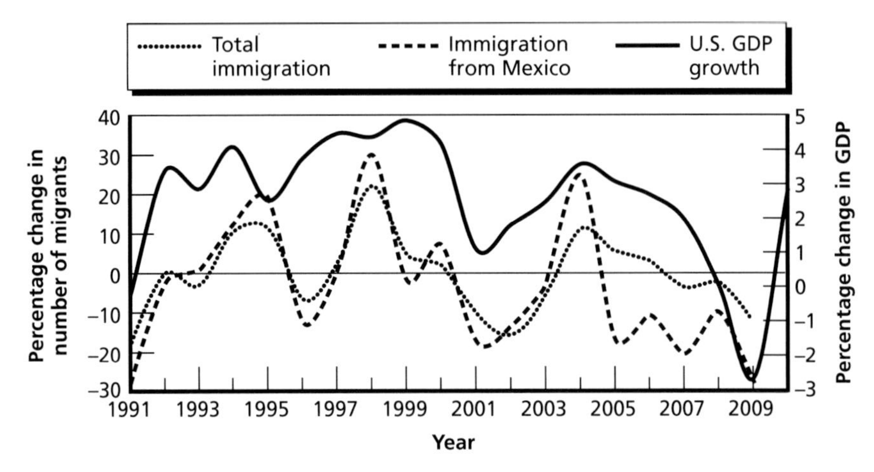

SOURCES: Passel and Suro, 2005; Pew Hispanic Center, 2007; BEA, 2010.
RAND *MG985/1-6.1*

the significant wage differentials between the two countries, shown in Table 6.1. For example, a Mexican between 23 and 27 years old with four years of education is likely to make almost six times as much in the United States as he or she would in Mexico. Those differentials decline with education and age.

It is worth mentioning that the wage differential is very similar for all age groups that have the lowest levels of education. According to the wage-differential hypothesis, the least skilled have more incentives to migrate to the United States.

Supply-Push: Performance of the Mexican Economy

Mexican Economic Conditions Are Critical to the Decision to Migrate

Immigration to the United States from Mexico has risen following economic problems in Mexico, such as the debt crisis in 1982 and the exchange-rate collapse in 1994 (Massey and Singer, 1995; Passel and Suro, 2005). Economic instability leads to unemployment and low real

Table 6.1
Ratio of U.S. Wages to Mexican Wages for Mexican-Born Workers, 2000

	Years of Schooling Completed					
Age	**4**	**5–8**	**9–11**	**12**	**13–15**	**16+**
18–22	5.8	4.9	4.2	3.9	3.4	2.2
23–27	5.9	4.6	3.9	3.2	2.5	2.5
28–32	5.3	4.4	3.6	3.0	2.0	2.4
33–37	5.7	4.4	3.6	2.9	2.2	2.4
38–42	5.6	4.4	3.2	2.9	2.2	2.2
43–47	5.8	3.9	3.1	2.4	2.2	2.0
48–52	5.8	4.1	3.0	2.2	1.9	2.0

SOURCE: Hanson, 2006.

NOTE: Mexican wages are rescaled to adjust for cost-of-living differences between Mexico and the United States, using the 2000 purchasing power parity (PPP) adjustment factor for Mexico, as listed in Hanson (2006).

wages. Figure 6.2 presents the real minimum wage from 1986 to 2010. In this figure, we observe that economic crises have eroded real wages in Mexico. Although the real minimum wage has been stable since 2000, the lack of job opportunities in Mexico has led to high migrant flows.

Unemployment in Mexico Is Relatively Low but Still Stimulates Migration

According to employment statistics, unemployment in Mexico is low when compared with other countries with similar development levels. However, these numbers can be misleading because Mexican unemployment statistics do not include some individuals who would be counted as unemployed under U.S. measurement standards. For instance, due to a weak unemployment-compensation scheme, persons without work in Mexico are often forced into marginal activities (e.g., street vending, moving, repairing), which results in their classification as employed rather than unemployed even if they work as little as one

Figure 6.2
Real Minimum Wage in Mexico, 1986–2010

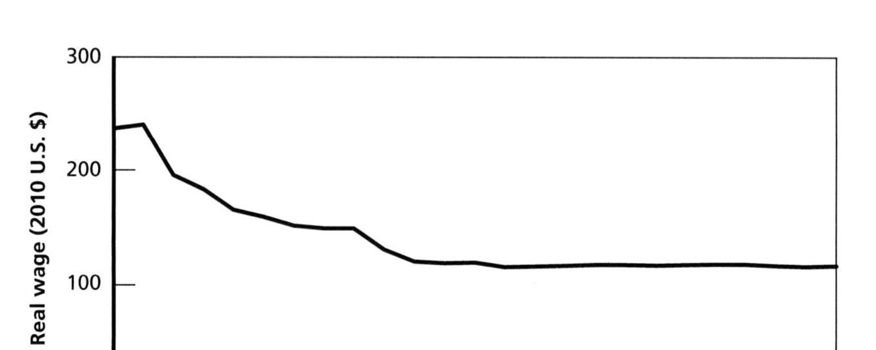

SOURCES: CONASAMI, 2009; BANXICO, undated.
RAND *MG985/1-6.2*

hour per week. According to Fleck and Sorrentino (1994), the reported rate would still be relatively low after its adjustment to the U.S. concept.

The last time the unemployment rate in Mexico was at a level considered high relative to rates in other countries (around 8 percent) was following the December 1994 peso devaluation. After achieving macroeconomic and financial stability, the private sector and temporary government public works programs absorbed a sizable fraction of the Mexican active population, which led to a boost in employment (Lustig, 1998).

Figure 6.3 shows a clear inverse relationship between migratory flows from Mexico to the United States and Mexican GDP growth during the 1990s. After 1999, however, Mexican GDP growth and migration flows appear to move in the same direction, an indication that both factors were following the U.S. GDP decline and recovery during those years (as observed in Figure 6.3). Figure 6.3 also shows the unemployment rate in Mexico, which fluctuates significantly less than migration flows except for the spike during the 1995 crisis but still appears to be related to them: Higher unemployment is associated with more migration to the United States.

Figure 6.3
Annual Change in Immigrant Flows from Mexico to the United States, Mexican Unemployment Rate, and Mexican and U.S. Gross Domestic Product, 1991–2010

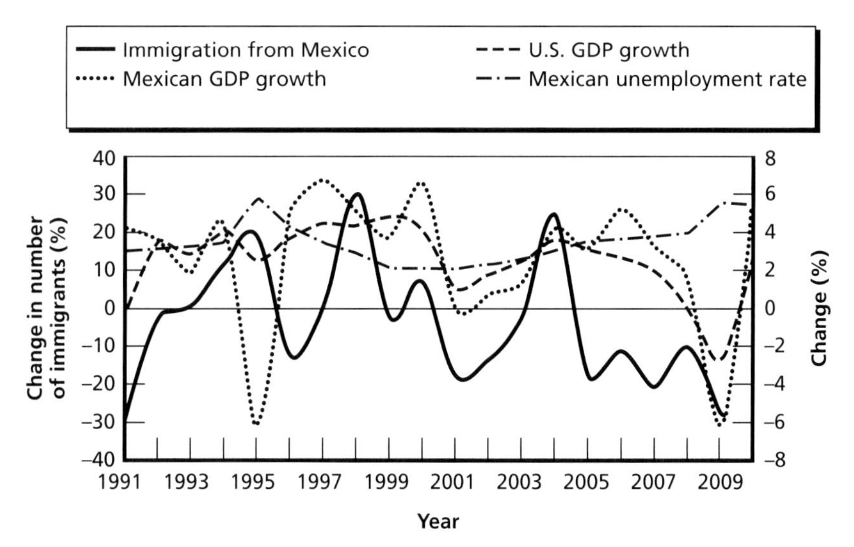

SOURCES: Passel and Suro, 2005; BANXICO, undated.
RAND *MG985/1-6.3*

The 2007 U.S. Recession Increased Unemployment in Mexico

Given that the Mexican economy grew more slowly in 2008 and contracted severely in 2009 due to the economic recession in the United States (and elsewhere), there was another rise in unemployment in Mexico. According to INEGI, Mexican annual real GDP grew 3.3 percent in 2007 and by an additional 1.5 percent in 2008 but fell by 6.1 percent in 2009. It rebounded to 5.5 percent in 2010. Unemployment rose until the third quarter of 2009, when it peaked at 6.2 percent, still lower than in the 1995 economic crisis. It had fallen to 5.3 percent as of the fourth quarter of 2010. The actual performance of the unemployment level depends on just how sluggishly the economy performs because economic stagnation naturally leads to less hiring and the loss of jobs. As explained earlier, the impact that higher unemployment has on migration flows is not entirely clear because U.S. economic growth appears to dominate the supply-push effects of Mexican unemployment. The Pew Hispanic Center estimates that the number of unauthorized immigrants from Mexico living in the United States did not increase from 2007 to 2008 (Passel and Cohn, 2008b).

Poverty and Economic Performance in Mexico

In 2010, Mexico was ranked 56th in the United Nations' Human Development Index, falling within the classification of high human development (UNDP, 2010). According to this measure, Mexico's main deficiencies are in the proportion of population that is illiterate, malnutrition, scarcity of physicians, and high income inequality (UNDP, 2007). Mexico's purchasing power–adjusted GDP per capita in 2010 was US$14,566.

We analyze poverty using two official measures reported by Mexico's government: (1) *capability poverty*, which refers to a person's inability to satisfy his or her minimum requirements of food, health, and education, and (b) *food poverty*, which is a person's incapacity to purchase the minimum necessary food basket, even if all his or her income were allocated to it. The 1994 economic crisis brought a significant increase in the incidence of poverty in Mexico, with capability poverty growing from 30 percent in 1994 to 46 percent in 1996 and food poverty increasing from 21 percent to 37 percent in the same period. However, as economic recovery began in 1996 and continued throughout the rest of the decade, poverty rates declined steadily, so that, by 2002, capability and food poverty had reached levels of 27 and 20 percent, respectively (see Figure 6.4). It took five years to observe a recovery from the 1995 economic crisis.

The economic crisis of the late 2000s reversed the decline in poverty at least temporarily. Between 2004 and 2005, there were no statistically significant changes in poverty rates then important reductions in poverty between 2005 and 2006 such that, by 2006, the rate of capability poverty had reached a value of 21 percent and only 14 percent of the population was living in food poverty, compared with their peak values of 46 and 37 percent, respectively. The 2010 figures reported by the National Council for Evaluation of Social Development Policy (Consejo Nacional de Evaluación de la Política de Desarrollo Social, or CONEVAL) (undated [b]), a Mexican govern-

Figure 6.4
Poverty Rates in Mexico, 1992–2010

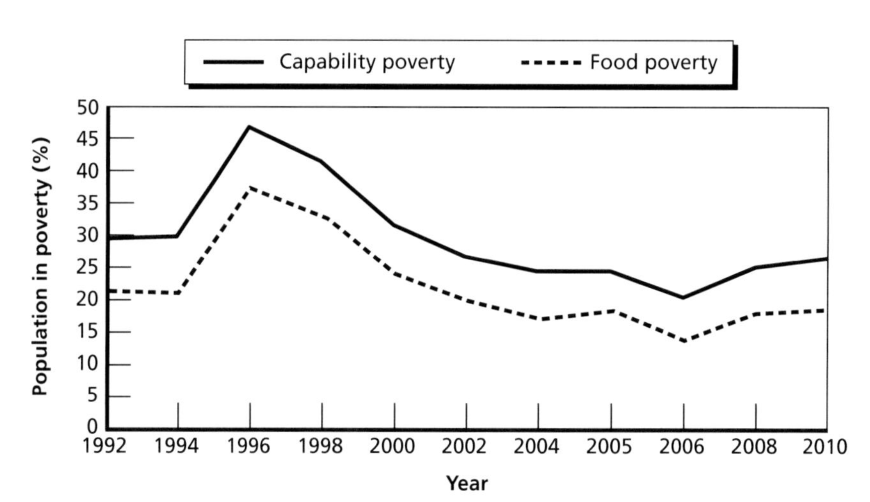

SOURCE: CONEVAL, undated (b).
RAND *MG985/1-6.4*

ment agency that evaluates social policies and measures poverty levels, show that food and capability poverty increased to 19 and 27 percent, respectively, effectively reversing the gains in the fight against poverty that had occurred from 2005 to 2006. Lower poverty rates do not seem necessarily to translate into reductions in migratory flows, however, as can be seen by comparing the migratory flow shown in Figure 6.3 and the poverty rate in Figure 6.4: The decreases in poverty that started in 1996 precede reductions in migratory flows from 1998 to 2001, but poverty continued to decrease up to 2004, and migration rose again between 2002 and 2004. Migratory flows declined substantially in 2005, and poverty slightly increased. From 2006 to 2007, migratory flows and poverty followed a negative correlation; in 2008, poverty and migratory flows both increased. In summary, we cannot observe a clear correlation between migratory flows and poverty.

Poverty Rates Differ According to Region

Average poverty figures might disguise the underlying differences between urban and rural communities. Poverty rates are significantly higher in *rural and small-town* communities—defined as those with fewer than 15,000 inhabitants. As of 2010, 29 percent of rural households lived in food poverty, but merely 13 percent of urban households were in that condition. Similarly, 38 percent of rural but only 20 percent of urban households lived in capability poverty in that year (CONEVAL, undated [a]).

Despite the larger incidence of poverty in rural communities than in urban ones, the first decade of the 21st century saw reductions in poverty levels only in rural, not urban, communities: The percentage of households in rural food poverty decreased by 12 percent, going from 50 to 38 percent in that period, and urban food poverty remained at 20 percent (after first dropping significantly to 14 percent in 2006). The declines in the percentage of households in capability poverty between 2000 and 2010 are similar to those in food poverty: 13-percent decline for rural communities and no decline for urban communities.[1]

In the past six decades, Mexico has undergone a rapid process of population movement toward urban areas. Between 1950 and 2010, the population living in rural areas decreased from 57 percent to only 22 percent of Mexico's total population. Despite this rural-urban population shift, rural regions still constitute more than 80 percent of the Mexican territory; thus, the bulk of Mexico's population concentrates in a few urban centers, and the rest is dispersed in a large number of small localities. Individuals from rural areas, though no longer constituting the majority of emigrants, are still overrepresented among the migrant flows (Riosmena and Massey, 2010).

The states with the highest levels of rural population also have the highest levels of poverty. The top panel in Figure 6.5 shows poverty levels for every municipality in the country, using a poverty index based on the 2005 Mexican census developed by CONAPO. Poverty is higher in the middle and southern regions of the country. The map also shows that regional inequality is extreme: We can see that most municipalities have either high/very high or low/very low poverty, while only a small number of municipalities are in the category of medium poverty.

---

[1] These figures vary, but only slightly, when persons instead of households are considered, since rural households are typically larger than urban households.

grown relative to temporary migration in recent years (García Zamora, 2005a). The recent increase in deportations and legal measures against illegal immigration at both the federal and local levels in the United States might also result in reduced migration or increased returning migrant flows, which would also reduce remittance flows to Mexico. In fact, both the number and dollar amount of remittances grew minimally between 2006 and 2007, and estimates by the Mexican Central Bank (Banco de México, or BANXICO) indicate that remittances in 2008 were nearly 4 percent lower than in 2007 and then dropped by 16 percent from 2008 to 2009 (BANXICO, undated). When we compare the first nine months of 2010 with the same period in 2009, however, it seems that the dollar amount of remittances has stabilized. Therefore, even if effective mechanisms are developed to maximize the positive impact of remittances on the regions that receive them, both families and governments need to develop strategies to prepare for the potential volatility in remittance flows.

Networks

In addition to demand-pull and supply-push factors, existing migrant networks can contribute to additional migration flows. For example, U.S. employers can use current employees to get word back to Mexico about job availability, and the presence of family members or people from the same village or neighborhood can help ease transition costs for the new migrants by providing them with financial assistance, lodging, food, or help finding employment.

As discussed in the previous section, most migration from Mexico to the United States has been from the central and western portions of the country. The importance of migration networks is evidenced by the persistence of these source regions: The correlation between sending states in the periods 1955–1959 and 1995–2000 is 0.73 (Hanson, 2006). In fact, Massey and García España (1987) argue that social networks help explain the surge of migration between 1970 and 1980, a period when real wages were falling and unemployment was high in the United States while wages were increasing and unemployment was low in Mexico.

Zuñiga Herrera, Leite, and Rosa Nava (2004) report that 80 percent of temporary migrants with family or friends in the United States have received some type of assistance. Amuedo-Dorantes and Mundra (2007) find that family and friendship ties raise unauthorized immigrant wages by an average of 2.6 and 5.4 percent, respectively, and those of legal immigrants by 8 and 3.6 percent. Higher wages might effectively increase the returns to migration, therefore providing a stimulus to continued emigration.

Munshi (2003), using data from well-established migrant-sending communities, finds that preexisting social ties of migrants ensure that those migrants receive various forms of assistance during their trips to the United States. These ties have an effect not only on location patterns of Mexican migrants but also on the occupations they choose, so they could explain the persistence of low-skill occupations among Mexican migrants.

CHAPTER SEVEN

Conclusions and Policy Recommendations

Despite the importance that Mexican immigration to the United States has for both countries, immigration policies in the two countries are not internally consistent and do not complement each other. Mexico does not officially encourage labor migration to the United States, but neither does it make serious efforts to discontinue the flow; the United States tries to put limits on Mexican immigration through policies, such as guest-worker programs, increased border enforcement, and restricting illegal immigrants from public benefits and services. Our findings suggest that, in the past, the implementation of stronger U.S. immigration policies has not had a large effect in deterring the flow of Mexican migrants. However, the impact of measures that were implemented in the 2000s cannot yet be determined.

Evidence suggests that the economic impact of immigration in the United States is modest. Our review of the literature suggests that Mexican emigrants to the United States place only a small additional social welfare burden on the United States by creating additional fiscal demands or labor-market distortions. However, emigration has important social and economic effects in Mexico, both positive and negative: The development of migrant social networks has resulted in a continuous and increasing process of migration, which has increased household income but also altered the composition of families.

National economic conditions greatly affect migratory decisions, and changes in personal labor situations prompt immigration decisions on both sides of the border. Both relevant stages of the life cycle—working life and retirement—are intermingled with the decision to migrate between the United States and Mexico. Retirement is one of the main motives to move from the United States to Mexico, and finding jobs is the principal reason for moving from Mexico to the United States. Other factors that are strongly correlated with the flow of migrants from Mexico to the United States are the performance of the U.S. and Mexican economies (as they are related to the supply of jobs in the United States and job opportunities in Mexico), large wage differentials between the two countries, and networks. These motives for migration vary in importance depending on the circumstances of potential migrants. Regional disparities and lack of development make it important to design different policies that can target such a heterogeneous population. Moreover, the migrant-sending regions in Mexico have spread throughout the country, and the flow of migrants might increasingly come from lower-income backgrounds and lower levels of skill than in previous decades.

Implications for Policy

Migration movements from Mexico to the United States are primarily motivated by economic factors. The recommendations in this section emerge from the analysis and conclusions of Part One of this monograph. By pursuing some or all of these actions, the United States and Mexico could reap rewards in terms of concrete improvement in their relationship, as well as better supporting U.S. border security and improving the general well-being of Mexicans legally immigrating to the United States.

Pursue a Legal Labor Market Driven by the Supply of U.S. Jobs

Building a strictly legal labor market will be key to enhancing the U.S.-Mexican relationship in the coming years. Mexico has a strong base of human capital, which frequently seeks out a broader base of economic activity through employment in the United States. Unfortunately, many immigrants enter illegally and are hired illegally, forming a kind of underground labor market in the United States. Reform thus should start with those sectors in which illegal immigrants are hired, and the system of issuing visas should be restructured and streamlined to better meet the needs of legal industries in need of importing labor.

Design and Implement Tools to Assist with Current Standards of Border Control

Immigration policy should be reexamined and designed to include the Mexican government's renewed commitment to effectively decrease illegal activities at the U.S. border. For example, databases that record individuals' labor histories in the United States and Mexico might be improved to help understand migration flows.

Design and Implement an Organization Committed to Making Labor Movement Transparent to Both Countries

The design of an organization to regulate and monitor labor movements between Mexico and the United States should have a binational approach, including input by officials from both countries, which would make all correlating data systems homogeneous at the implementation level. This organization would keep and have access to accurate databases of each individual's previous work history, employers, qualifications, occupation, employment and unemployment periods, contributions to the social security systems in the United States and Mexico, and family characteristics of Mexican workers who are or have been employed in the United States. This organization could also help to facilitate a U.S.-Mexican social security totalization agreement, which would make possible the portability of social security benefits between the two countries.[1]

Although such a binational organization might seem to arise from a futuristic view of U.S.-Mexican relations, it nonetheless has the capability of fulfilling specific needs of both nations. After the terrorist attacks in the United States in 2001, the United States has renewed its interest in preventing the illegal entry of people and goods coming in from its southern border. Mexico, in turn, must be urged to realize the important potential for economic growth and stability from

---

[1] For more information on a U.S.-Mexican social security totalization agreement, see, for example, Aguila and Zissimopoulos (2010).

remittances. Developing a binational organization committed to international security and improving the labor force is one effective solution, but it requires a strong commitment and collaboration from both countries.

Begin to Address U.S. Immigration from the Other Side

Many current policies address immigration from the perspective of U.S. business investment or are aimed at controlling migration flows from specific sending regions. It is important that policymakers from both countries begin to understand the causes and consequences of Mexican migration, such as states of origin of migrants, regional disparities, and the economic and social situation in Mexico. Through such understanding, policies that assist and promote employment in underdeveloped areas will help retain population in its places of origin. A proportion of the population will still migrate from Mexico to the United States due to networks or cultural tradition to move north. However, a thorough analysis aimed at understanding characteristics of migrants who are likely to stay in the place of origin will help policymakers design appropriate economic incentives.

PART TWO

Progress and Challenges: Mexico's Economic and Social Policy

It is your concern when your neighbor's wall is on fire.
—*Horace (Epistles, book 1, epistle xviii, l.84)*

In this second part of this monograph, we examine Mexico's progress and challenges on the economic and social fronts. Although the repercussions of the U.S. recession that started in 2007 and the global crisis that commenced in 2008 are clearly of great interest to both countries, we focus our analysis predominantly on the years preceding the recession and global crisis in order to concentrate on the secular trends we have witnessed during the past two decades.

Although Mexico's economic performance was relatively stable before the 2007 U.S. recession, it continues to be classified as a middle-income country. Its performance has been far from stellar. As the World Bank (2006a, p. ix) notes,

> While real GDP per capita in Mexico only grew at an annual rate of 1.2 percent between 1994 and 2004, it grew at 7.7 percent in China, and at 5.9 percent on average in the East Asia and Pacific region. Productivity growth rates have also been lackluster.

Performance on the social front has likewise been mixed. Growth has been accompanied by increased inequality, and poverty is still widespread, especially in rural areas and among the indigenous people. Provision of basic education has improved, but the quality of the provided education is suspect. Access to education, especially higher education, is limited. There is a plethora of public social programs, and the leading light on this front, Oportunidades, has been heavily evaluated and widely imitated elsewhere in Latin America.[1] But there are many others that have not been formally evaluated and whose performance is unknown. Social security has been reformed, but its reach, similar to that of health provision, is highly unequal.

On both the economic and social fronts, it is clear that, although Mexico has made strides, there is an urgent need for reforms on several fronts.

Understanding the economic and social situation in Mexico is important for understanding the policy challenges Mexico faces. It also has implications for U.S. immigration and trade. As we saw in Part One, lackluster performance of the Mexican economy increases the likelihood of immigration from Mexico to the United States. It could also have a negative effect on bilateral trade between the two countries; Mexico is, after all, the third-largest trading partner of the United States after Canada and China (Koncz-Bruner and Flatness, 2011).

We begin our discussion with Chapter Eight, in which we describe the economic and demographic landscape in Mexico. Chapter Nine examines the results of specific economic reforms more closely, and Chapter Ten provides the outcomes of important social reforms made since the 1990s. Finally, the concluding chapter of this section identifies for U.S. and Mexican policymakers some potential targets of programs and policies to support and strengthen Mexican economic and social reforms. For additional background on economic and social conditions and policy in Mexico, we summarize Mexico's political history in the appendix.

---

[1] As we discuss in Chapter Ten, later in Part Two, Oportunidades is a conditional cash-transfer program (families receive monetary transfers conditional on their fulfilling their obligations) that integrates education, health, and nutrition interventions.

CHAPTER EIGHT

The Economic and Social Landscape of Mexico

In this chapter, we provide an overview of the economic and social landscape of Mexico and, in the next two sections, delve into greater detail in these areas in order to identify policy challenges.

Mexico's Economy: Crisis and Recovery

Beginning in the mid-1980s and accelerating in the early 1990s, Mexico began the process of recovering from a series of economic crises it experienced in the 1970s and 1980s. The years 1976, 1982, and 1987 were especially bad. Inflation skyrocketed to 98.8 percent in 1982 and 159.2 percent in 1987. The 1982 crisis was attributed to current account and budget deficits. A drop in oil prices of 50 percent during 1986, which substantially decreased government revenues, precipitated the 1987 crisis. Mexico has had a strong dependence on oil revenues, which we discuss later. The drop in the main international stock exchanges worsened the crisis (Medina Peña, 1995).

The economic recovery of the 1990s was based on decreasing the size of the government, privatizing government companies, reducing the value-added tax (VAT) for most goods, and decreasing the income tax. These actions and the creation of NAFTA in 1994 ushered in expectations of a sustainable economic path in Mexico. Table 8.1 highlights some of the major economic reforms enacted in Mexico since 1988. As seen in Figure 8.1, real GDP growth stabilized around 3.5 percent; as shown in Figure 8.2, inflation dropped to below 10 percent during the beginning of the 1990s. Unemployment, which exceeded 5 percent in

Table 8.1
Major Mexican Economic Reforms Since 1988

Date	President of Mexico	Reform
1988–1994	Carlos Salinas de Gortari	Deregulation undertaken and the role of the government is decreased; the private sector is promoted.
December 22, 1994	Ernesto Zedillo Ponce de León	A flexible exchange rate is adopted for the Mexican peso.
1997	Ernesto Zedillo Ponce de León	PAYG social security system (IMSS) is replaced by PRAs for workers in the private sector.
2004	Vicente Fox Queseda	BANXICO is made independent.
2007	Felipe Calderón Hinojosa	PAYG social security system (ISSSTE) is replaced by PRAs for workers in the public sector.

SOURCES: Medina Peña, 1995; IMF, 2006b; Aguila et al., 2011.

NOTE: PAYG = pay as you go. IMSS = Instituto Mexicano del Seguro Social, or Mexican Social Security Institute. PRA = personal retirement account. ISSSTE = Instituto de Seguridad y Servicios Sociales de los Trabajadores del Estado, or State Workers Security and Social Services Institute.

the mid-1990s, decreased and stabilized through mid-2008, the onset of the global financial crisis (see Figure 8.3). Less economic uncertainty during the beginning of the 1990s led to a period of low interest rates (see Figure 8.4). An increase in stability, a decrease in import duties, and an increase in openness in general boosted foreign investment considerably at the beginning of the 1990s (see Figure 8.5).

On December 19, 1994, an unexpected economic crisis started with the depreciation of the Mexican peso followed by a generalized outflow of capital (see Figure 8.5). In particular, investors sold Ajustabonos, government bonds linked to the exchange rate, and Tesobonos, government bonds issued in U.S. dollars (Sánchez Daza, 2004). GDP shrank by 6.2 percent between 1994 and 1995, and inflation hit 35 percent (see Figures 8.1 and 8.2). This was a major economic crisis, and recovery took approximately five years. Although a

Figure 8.1
United States and Mexico Real Gross Domestic Product Growth, 1990–2010

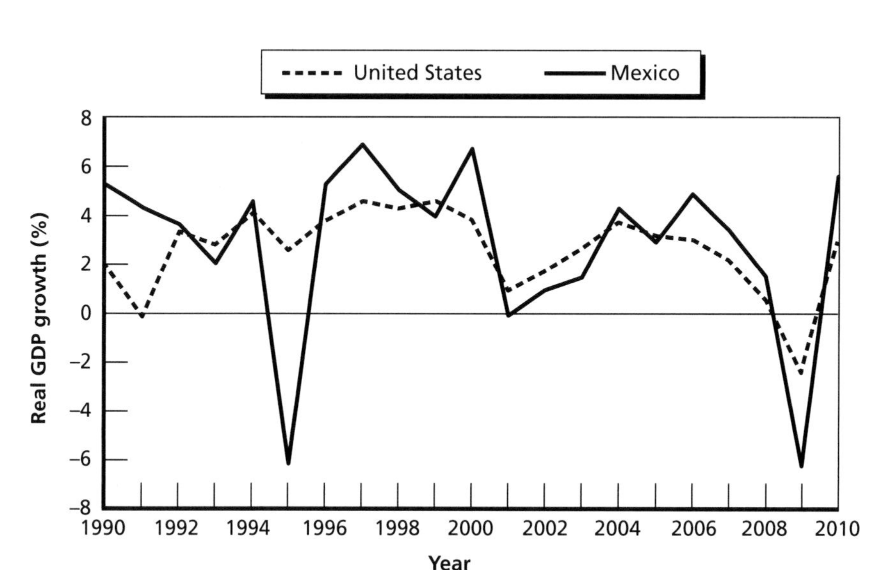

RAND *MG985/1-8.1*

fixed exchange rate can provide stability during normal times, it can precipitate currency crises when there are shocks to the economy. The Mexican peso was floated on December 22, 1994 (see Figure 8.6), and depreciated relative to the U.S. dollar over the next few years. In general, for most of the past decade, the exchange rate was stable (see Figure 8.6); however, between July 2008 and February 2009, the Mexican peso depreciated by 50 percent. In the three months following this period, the Mexican peso regained some strength and appreciated by 12 percent. As of March 31, 2011, the exchange rate was 11.9 Mexican pesos per U.S. dollar.

Inflation started to decrease sharply in 1996, and GDP growth returned to stability between 1996 and 2000 (see Figures 8.1 and 8.2). Figure 8.4 shows that interest rates were highly volatile during the second half of the 1990s. Foreign investment grew at an average annual rate of 12.5 percent between 1996 and 2000. Remittances grew annually 15 percent on average, and income from tourism grew 7.7 percent during this period.

Figure 8.2
United States and Mexico Inflation, 1990–2010

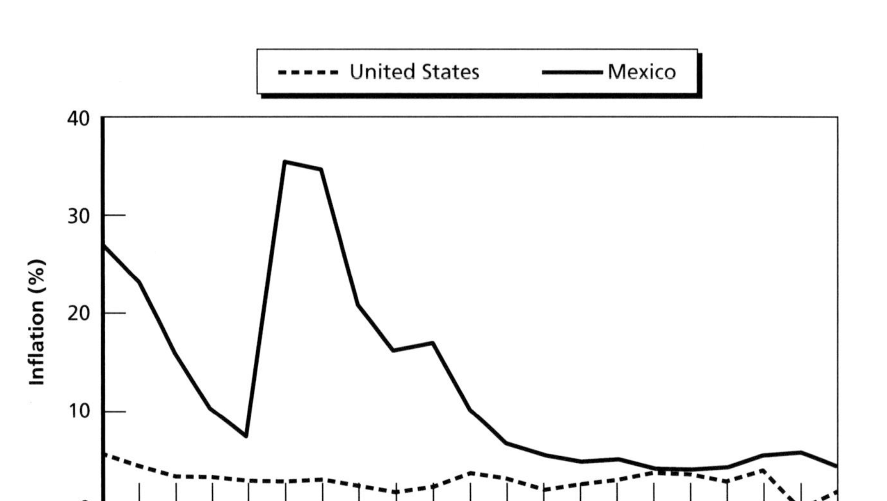

SOURCES: INEGI, undated (a); U.S. Census Bureau, 2011b.
RAND *MG985/1-8.2*

A major social security reform was implemented in 1997 while President Zedillo was in office. Following the Chilean reform, PRAs managed by private banks substituted for the traditional PAYG system, mainly for private-sector workers. The main public social security system did not move from a traditional PAYG system to PRAs until April 2007. One of the main goals of the social security reform was to increase private savings by introducing pension funds, managed by the government and transferred from young generations to old generations, to the financial market (Aguila, 2011).

There have been encouraging signs on the Mexican macroeconomic front since 2000. As seen in Figure 8.2, Mexican inflation, which was 16.6 percent in 1999, dropped to 3.6 percent in 2006, approaching U.S. levels. Up to the U.S. recession that began in 2007, GDP growth was moderate, fluctuating around 2.5 percent. The GDP growth in Mexico between 2007 and 2010 had a trend similar to that of the United States. However, there was a sharper decline in GDP growth

Figure 8.3
Quarterly Unemployment Rate as a Proportion of the Economically Active Population in Mexico, 1990–2010

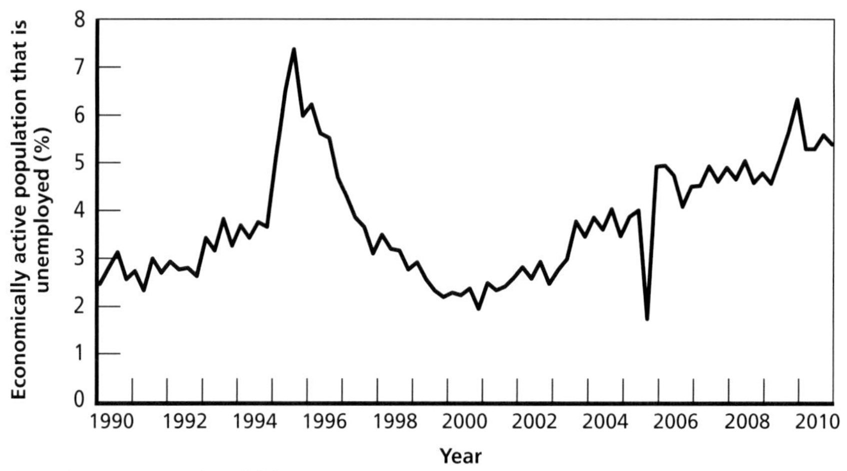

SOURCE: INEGI, undated (b).
NOTE: This figure differs from the annual unemployment rate presented in Figure 6.3 in Chapter Six because these data are obtained from monthly reports showing a higher quarterly variability.
RAND *MG985/1-8.3*

in Mexico in 2009. Figure 8.4 shows a drop in the Mexican interest rates (Federal Treasury Certificates, or Certificados de la Tesorería de la Federación [CETES]) since 2000, again approaching U.S. rates for prime lending. Firms started issuing long-term bonds, and Mexico has been moving toward a better-developed financial sector. In Figure 8.6, we observe a diminished tendency for the Mexican peso to depreciate relative to the U.S. dollar from 2000 until the onset of the global financial crisis. Since 2008, the Mexican peso again depreciated relative to the U.S. dollar. Figure 8.5 shows a sharp increase in remittances since 2000. As reported by BANXICO, remittances, which were $6.6 billion in 2000, increased at an annual average rate of 23 percent until 2007, at which time they reached $26.1 billion. As a result of the economic crisis, growth rates became negative in 2008 and 2009, with remittances falling to $21.2 billion in 2009.

Figure 8.4
U.S. and Mexican Interest Rates, 1990–2010

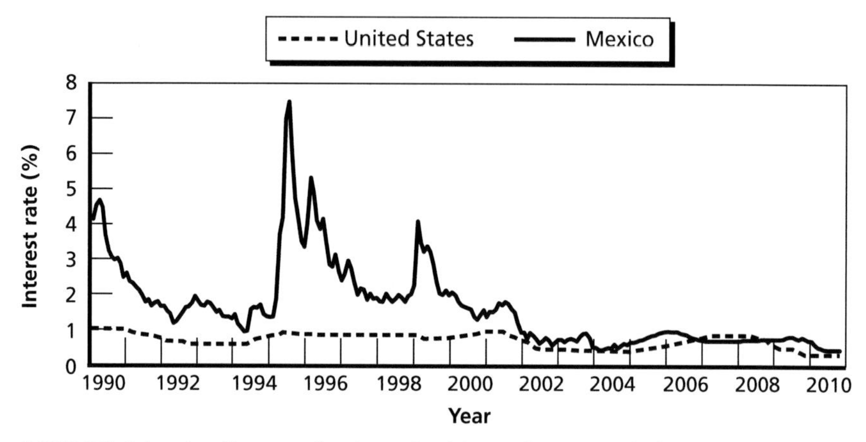

SOURCES: Prime lending rate for the United States from Board of Governors of the Federal Reserve System (undated); CETES for Mexico from BANXICO (undated).
RAND *MG985/1-8.4*

Figure 8.5
Mexican Sources of Foreign Exchange, 1990–2010

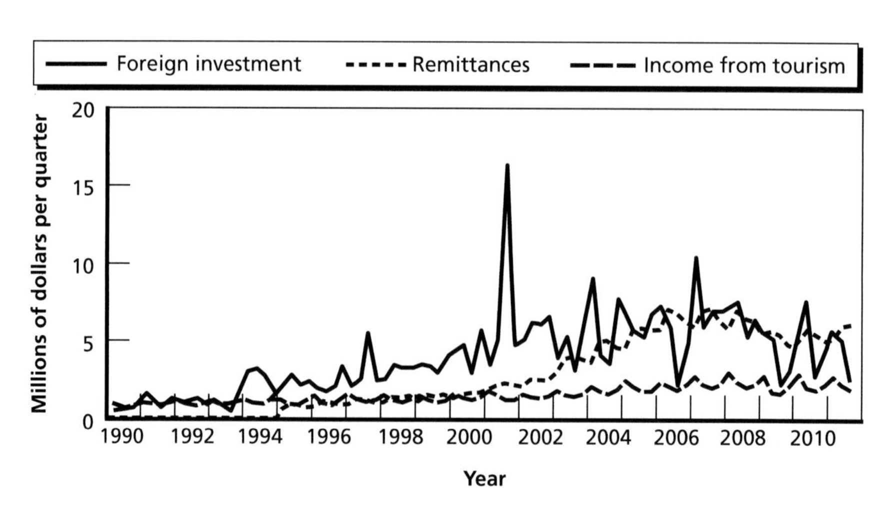

SOURCE: BANXICO, undated.
RAND *MG985/1-8.5*

Figure 8.6
U.S. Dollar–Mexican Peso Exchange Rate, 1990–2010

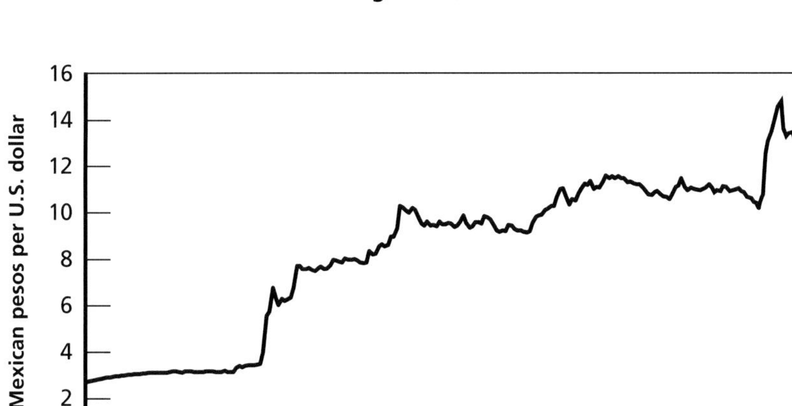

SOURCE: BANXICO, undated.
RAND *MG985/1-8.6*

Increasing oil prices since 2000 appear to have, in part, contributed to Mexico's macroeconomic stability up to 2008. Figure 8.7 presents the price of Mexican and Brent crude oil (a major classification of oil, traded in international markets) since 1990. The price for Mexican crude oil was US$22.9 per barrel in January 2000 and peaked at US$122.4 in July 2008. It then dropped precipitously but then again embarked on a steady upward climb. The Mexican government became increasingly dependent on oil revenue as oil prices rose. In 1990, oil revenues corresponded to 30.1 percent of total government revenues, and this rose to more than 35 percent from 2004 to 2008, peaking at 38.0 percent in 2006, before falling back to 31.0 percent in 2009. In 2010, it rose again, to 32.9 percent. Later in this section, we discuss the potential pitfalls of this dependence and how the contribution to the economic health of Mexico could be short-lived.

Figure 8.7
Mexico and Brent Crude Oil Prices, 1990–2010

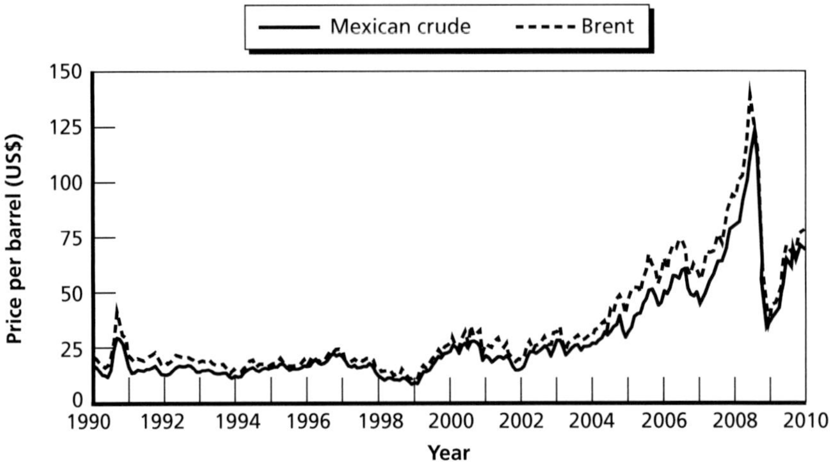

SOURCES: Swiss National Bank, 2011; BANXICO, undated.
RAND *MG985/1-8.7*

Mexico's Social Policy: Evolving and Improving

As with economic policy, social policy in Mexico has improved in the past several decades. Mexico's successful poverty-alleviation program, Oportunidades, has been imitated in many other countries. Social policies in other areas, including education, health-care access, and social security provision, have improved the well-being of the population.[1] However, there are still many deficiencies generating inequality and poverty that need to be tackled in order to improve productivity and further promote the population's well-being.

The demographic structure of Mexico has been a major driver of these social policies. Figure 8.8 presents the structure of the population in 1950, substantial growth in the following five decades reflected in the corresponding structure for 2005, and the process of aging reflected in future projections for 2050.

---

[1] We cover these in greater detail in Chapter Ten.

Figure 8.8
Mexican Population, by Age Group, 1950, 2005, and 2050

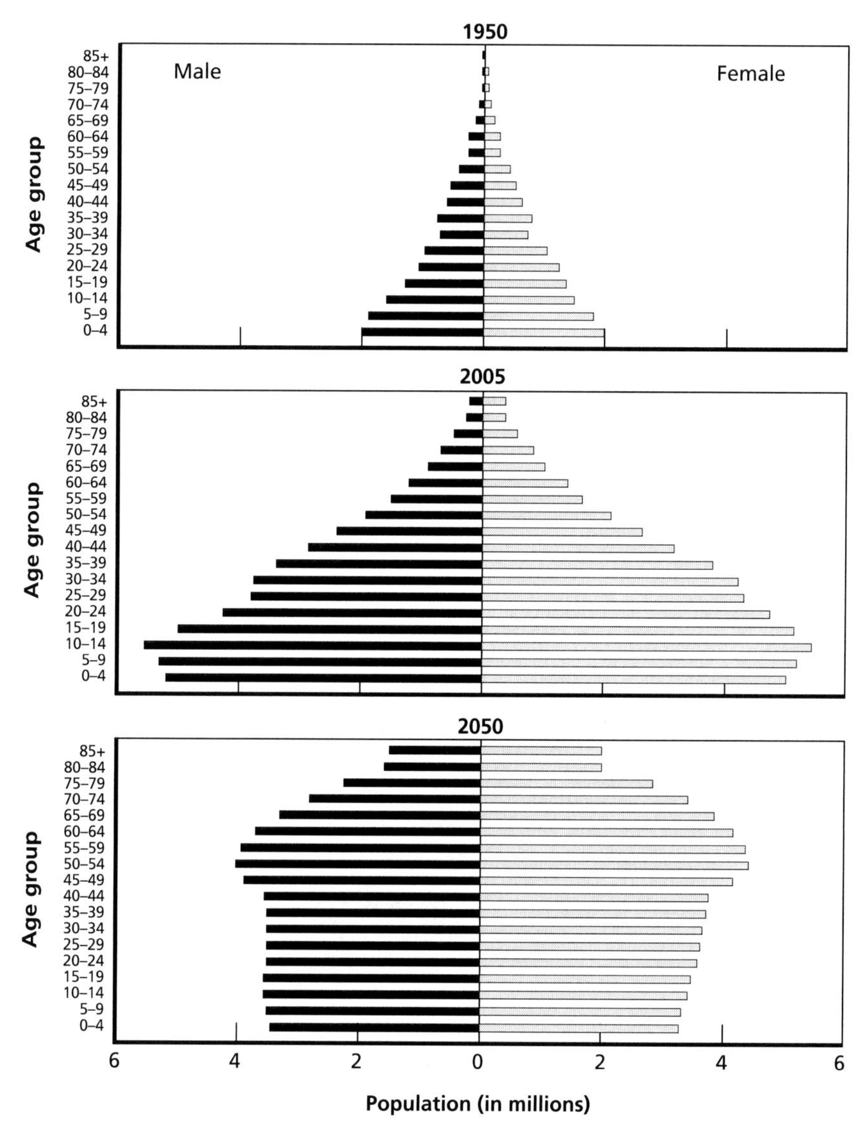

SOURCES: INEGI (1950, 2005a); CONAPO (2006).

RAND *MG985/1-8.8*

The total population in 1950 was 25.7 million, and most of the population was 14 years old or younger. During that period, the highest priorities in social policy were primary and secondary education and maternal and child health care. From 1950 to 2005, increased fertility (until the mid-1970s) and life expectancy caused the population to increase by 300 percent.

In 2005, the Mexican population had increased to 103.2 million, and the largest proportion of the population was 19 years old or younger. The fertility rate decreased from 5.7 children per family in 1976 to 2.2 in 2005. Individuals 65 years or older represented 5.5 percent of the total population in 2005, widening the top of the population pyramid. This rapid growth in population has increased the need for sound social policies on education and health care to increase productivity and generate growth. Without such policies, Mexico will not be able to realize the potential demographic dividend to be had in the next 25 years, when most of the population will be of working age. Meanwhile, the increase in the elderly population poses its own challenges. In addition to infectious diseases, attention needs to be devoted to chronic conditions, which are costly because treatments are given for longer periods of time. The lack of universal social security coverage also becomes an important consideration for the currently old, as well as for the wave of retirees to come.

The baby-boom generation in Mexico was born between 1980 and 2005, and, as the demographic projection in Figure 8.8 shows, individuals 65 years and older will represent 21 percent of the total population in 2050. The largest proportion of the population will be between 45 and 69 years old, at or close to the retirement age. Mexico will therefore face even stronger pressure from the social security systems and health-care provision for the middle-aged and elderly population in the future.

Mexico has a window of opportunity in the next 20 years for economic growth as the baby-boom generation enters the labor market. Mexico could experience an increase in remittances, but, when these generations start returning to Mexico around retirement age with low levels of wealth accumulation and lack of social security coverage, the health-care system and welfare programs will have to play major roles in securing the well-being of these generations. The burden will lie on future generations with fewer chances of success in a weak institutional environment and a poorly performing economy. For the United States, this scenario will also imply pressure on the economy. The United States, due to its proximity to Mexico, will share the burden of the effects of the demographic transition in Mexico. In sum, the economic and demographic trends of Mexico—in particular, its large proportion of working-age population—have significant implications for the United States. As discussed in Part One, lack of job opportunities and persistent wage differentials could increase the flow of immigrants.

CHAPTER NINE

Mexico's Economic Management

The economic recovery of the 1990s was spurred by a series of reforms: decreasing the size of the government, privatizing government companies, and moderating increases in public-sector wages.

The government privatized approximately 900 firms between the end of the 1980s and the beginning of the 1990s, among them commercial banks and airline, mining, and telecommunication companies. This allowed the government to obtain substantial revenues while decreasing its operational costs because many of the erstwhile public-sector firms had cash-flow deficits (Aspe Armella, 1993).

The past two decades have seen pro-market reforms aimed at increasing the competitiveness of many emerging economies. The aim has been to eliminate monopolistic practices, aiding consumers through competitive pricing and using market prices as signals to allocate resources efficiently. In other words, competition among firms has been seen as a route to increased national competitiveness through improvements in productivity and innovation. State-run enterprises in these economies have been privatized as part of the process. Mexico was part of this move toward economic liberalization, taking many important steps at the end of the 1980s and beginning of the 1990s.

In this chapter, we examine Mexico's key economic policies as they relate to the important areas of competition and privatization, taxation, labor, energy, and fiscal federalism. It is our goal here to clearly identify areas of progress, as well as challenges that remain. We also examine policy changes that might have contributed to Mexico's relative financial stability and improved U.S.-Mexican economic ties.

The Mexican Financial Crisis: Before and After

The Mexico of the 21st century is very different from the Mexico of the early 1990s, in great part because of a financial crisis in 1994 and 1995. Mexico is on more-solid ground today than it was prior to the crisis. Thanks to prudent macroeconomic policies, inflation and the public debt ratio are low, and the country enjoys both domestic and external economic stability. Mexico introduced a fully flexible exchange-rate regime, significantly reduced its current account deficit, and greatly strengthened financial regulation and supervision, all of which have played a role in Mexico becoming increasingly resilient to external shocks.

This crisis was the last serious economic crisis Mexico faced, at least through 2011; because it affected subsequent economic policymaking, we focus in this section on its causes and consequences. Table 9.1 presents selected economic indicators before the 1994–1995 crisis and before the onset of the 2007 U.S. recession. Keeping in line with our focus on longer-term structural trends, we present and discuss data up to 2006 (the full year for which data are available before 2007, start of the U.S. recession that led into the global crisis in 2008). Unlike the 1994–1995 crisis, the 2007–2008 economic crisis is global and not of Mexico's making. Presenting and discussing more-recent data will conflate the earlier crisis (which is of primary interest here) with the later crisis.

Primary Causes of the Crisis

The main cause behind the Mexican 1994–1995 crisis was an unsustainable current account deficit that, starting in 1990, was financed by very large capital inflows. Unprecedented amounts of capital flowed into the country, reaching US$104 billion between 1990 and 1994, which was around 20 percent of total capital flows to developing economies during that period. In the aftermath of the crisis, the current account deficit, which reached 8 percent of GDP in 1994, was reduced to 0.2 percent of GDP by 2006, and international reserves increased by more than US$60 billion between 1994 and 2006.

Table 9.1
Mexico's Economic Indicators, Before the 1994–1995 Crisis and 2003–2008

	Before the 1994–1995 Crisis		2003–2008	
Indicator	**Date**	**Value**	**Date**	**Value**
External current account deficit (as percentage of GDP)	1994	8	2006	0.2
BANXICO net international reserves (US$ billion)	1994	4.9	2006	67.7
Gross international reserves in months of imports of goods and services	1994	0.7	2006	3.0
Gross international reserves in percentage of net public-sector external debt	1994	6	2003	132.8
Gross external debt (as percentage of GDP)	1995	57.7	2006	20.1
Average maturity of external public debt, in months	1994	9	2006	36
Gross national saving (as percentage of GDP)	1994	15	2006	21.8
Crude oil price ($ per barrel)	1994	13	May 2008	135
Credit to private sector (as percentage of GDP)	1994	50.7	2006	22.1
Commercial banks' nonperforming loans (as percentage of total loans)	1994	12	2006	1.7
Exchange-rate regime	1994	Pegged	Since December 22, 1994	Floating

SOURCES: IMF, 2004, 2006a, 2007a, 2007c; World Bank, undated (c); Quintin and López, 2006.

Prior to the crisis, Mexico was borrowing excessively in order to finance consumption rather than capital investment. By 2006, Mexico did not rely on external borrowing the way it had more than a decade before. In addition, by 2006, Mexico was borrowing more for the longer term than it had ten years earlier, suggesting that it was borrowing to finance capital investment projects with longer maturity rather than consumption. The average maturity of Mexico's public debt as of 2011 was approximately 30 years, compared with barely nine months at the onset of the 1994–1995 crisis (Secretaría de Hacienda y Crédito Público [SHCP], 2011). The rate of domestic saving increased from 15 percent of GDP in 1994 to 21.8 percent in 2006.

After the Crisis

Banks and Banking. By 2006, the banking sector was in better shape than it had been in the early 1990s, largely because supervision greatly improved. The ratio of nonperforming loans was tiny compared with what it was in the early 1990s as a result of more-careful screening for creditworthiness of clients. BANXICO maintained strict monetary discipline backed by sound fiscal policy. The government kept budget deficits under 2 percent of GDP—with the budget even in surplus at times—through 2008, with both the overall public sector and federal government deficits exceeding 2 percent only in 2009 and 2010. Because of the government's monetary and fiscal policy, inflation became more stable.

Exchange Rate. Mexico's exchange-rate policy also contributed to its stability. In the past, attempts to maintain a fixed exchange rate resulted in balance-of-payments crises as the government ran out of foreign reserves while attempting to defend the peso. However, the exchange rate has been floating freely for more than a decade now. Also, starting in May 2008, the Mexican peso has been included as a settlement currency in continuous linked settlements (CLSs) along with 16 other major currencies, allowing it to be accepted by banks around the world as a form of payment and for exchange. This can be viewed as another indication of Mexico's financial stability.

Output and Productivity. Despite these achievements, output and productivity growth remained low in Mexico. Between 1994 and 2006, real GDP per capita in Mexico grew at an annual rate of only 1.2 percent. In contrast, Mexico's current account balance strengthened through 2006 thanks to rising oil exports and fast-growing remittances. Mexico still faces a variety of other economic challenges, to which we now turn.

Business Regulation and National Competitiveness

One element of increasing competitiveness is ensuring that the extent of business regulation is reasonable. Therefore, we start by examining the indicators of competitiveness in the Mexican economy. Commonly used measures show Mexico's performance to be mixed. The World Bank's *Doing Business* database aggregates competitiveness along several dimensions—starting a business, dealing with licenses, employing workers, registering property, getting credit, protecting investors, paying taxes, trading across borders, enforcing contracts, and closing a business—into a single index that captures the ease of doing business in a particular country.

In the 2011 rankings, Mexico stood 35th out of 183 countries, the highest ranking of any Latin American country (World Bank, undated [b]). It was 41st the year before. The improved ranking reflects changes in rules and procedures for starting a business, getting construction permits, closing a business, and getting electricity. Areas for improvement include registering property, paying taxes, and enforcing contracts.

In contrast, the 2011 *Global Competitiveness Index* (GCI) by the World Economic Forum ranks Mexico 58th worldwide. China, India, and Chile are some of the countries that compete with Mexico for foreign investment, and these countries had better placement in these rankings. Problem areas included institutions, goods market and labor-market efficiency, and financial market development. The 2011 *Index of Economic Freedom* ranks Mexico 48th out of 179 ranked countries (four were unranked) and classifies it as a "moderately free" economy (Miller et al., 2011). It ranked 41st in the 2010 survey. Weaknesses in 2011 included property rights, freedom from corruption, financial freedom, and labor-market rigidity.

Competition Across Mexican States

National rankings also mask the large disparities in competitiveness seen across the Mexican states. Understanding regional differences is particularly important because state and local governments control policies regarding procedures to open businesses. Table 9.2 captures these disparities. For each criterion, we note the states that perform the best and give their global ranking, as well as the average value for Mexico.

The states of Aguascalientes and Zacatecas (which neighbor each other) perform well on many of the dimensions. Even more striking is the *degree* to which their performance exceeds the average Mexican performance. For instance, the number of days it takes to register property in Zacatecas is less than one-fourth of the national average. Although such disparity might be disheartening in one respect, the better-ranking states prove that reforms are viable in Mexico and provide models for other states to emulate.

Table 9.2
State-Level Disparities in Doing Business in Mexico in 2008

Criterion	Global Ranking (175 economies)	Mexico's Best Performer	Mexico's Average
Days to open a business	34	Aguascalientes, Guanajuato, Puebla, Coahuila (12 days)	28 days
Cost to open a business	54	Campeche (7.4% of per capita GDP)	12.5% of per capita GDP
Days to register property	28	Michoacán, Zacatecas (15 days)	74 days
Cost to register a property	23	Aguascalientes (0.84% of property value)	4.8% of property value
Days to enforce a contract	13	Zacatecas (248 days)	415 days
Cost to enforce a contract	46	Michoacán (19.5% of debt)	32% of claim

SOURCE: World Bank, undated (b).

The higher regulatory burden captured by these indicators has a real economic impact. The Instituto Mexicano para la Competitividad (IMCO, or Mexican Institute for Competitiveness) (2005) estimates that regulatory burden costs Mexico 15 percent of its GDP, compared with 12 percent in Canada (L. Jones and Graf, 2001) and 8 percent in the United States (Crain and Hopkins, 2001).

Monopolies

Monopolies, both government and private, are cited in a survey of firms to be the largest obstacles to business development (Centro de Estudios Económicos del Sector Privado [CEESP], or Private Sector's Center for Economic Studies, 2005). Although the Mexican Constitution of 1917 prohibits monopolies and monopolistic practices, implementation of competition policies is relatively recent (World Bank, 2006a).

Policies and Regulations

In 1993, Mexico enacted the Federal Law of Economic Competition (Ley Federal de Competencia Económica, or LFCE) and established a modern competition policy administered by the Comisión Federal de Competencia México (CFC, or Federal Competition Commission). The CFC is an autonomous agency within the federal government, attached to the Secretariat of the Economy for budgetary purposes. The competition law that defines the responsibilities for the CFC is fairly comprehensive, prohibiting "absolute" monopolistic practices (such as price fixing and distribution restrictions). "Relative" monopolistic practices (such as restrictions on resale and exclusivity contracts) can be found illegal only if the enterprises employing these practices have sufficient market power and cannot provide an efficiency defense. The LFCE allows for investigation of concentrations, including mergers and acquisitions, by the CFC. Multiple reforms in specific industrial sectors followed the creation of the CFC, with regulatory agencies set up to work with the CFC on matters related to competition. The CFC's role in granting licenses and permits is outlined in regulatory laws for various sectors and activities, such as telecommunications, natural gas, and aviation (Ramsey, 2003; Shaffer, 2004).

In 2000, Mexico created the Federal Regulatory Improvement Commission (Comisión Federal de Mejora Regulatoria, or COFEMER) to improve regulation by promoting those regulations that produce greater benefits than costs and to ensure transparency of the regulatory process. COFEMER reviews regulatory impact assessments (RIAs), which need to be filed for all new regulations that impose costs on citizens.

In 2003, in response to moderating growth and increasing global competition, the Fox administration declared that increasing Mexico's competitiveness would be a key priority of the Mexican government.

In his survey of competition policy in Mexico, Ramsey (2003) concluded that, despite outpacing its Latin American neighbors in reforms, Mexico continued to be hampered by inefficiencies stemming from constitutionally sanctioned monopolies or poorly privatized industries. Because the CFC did not have the power to dissolve existing monopolies, it had little effective power other than to confront specific monopolistic practices. Given legal limitations, its enforcement power was weak, collecting only 10 percent of the fines it had imposed since its creation. (Under the law current at the time, the CFC did not directly collect fines, which accrue to a general treasury fund.) Ramsey argued for a more autonomous and powerful CFC with increased and higher-quality staff, its budget allocated by the Mexican Congress instead of the executive.

The Organisation for Economic Co-Operation and Development (OECD) peer review of the Mexican competition law and policy was more positive about the CFC. It stated that the CFC was willing to engage powerful economic interests and "has matured into a credible and well-respected agency that has compiled a remarkable record of achievement regarding the difficulties in its environment" (Shaffer, 2004, p. 7). However, it noted that the general support for competition policy was an open question, as was the CFC's ability to effectively and efficiently address anticompetitive conditions, especially because of certain deficiencies in statutory authority, its judicial review procedures, and methods of interface with other government entities. Moreover, the CFC had suffered a decline in resources even though its workload had increased.

Finally, in May 2011, Mexico completed a major reform of its competition law, effective May 11 (Harrup, 2011; "Mexican Competition Law," 2011). Among other reforms, the law gave the CFC new powers of investigation and injunction, increased the level of fines the CFC could levy, and criminalized some monopolistic practices. One of the investigative innovations was that the law allows the CFC to carry out unannounced visits to gather information; previously, it had to announce office visits in advance (OECD, 2011a).

Tax Policy

The tax-to-GDP ratio in Mexico is one of the lowest in Latin America. Including social security contributions and all payments by Petróleos Mexicanos (PEMEX) to the government, tax revenue amounted to 21.0 percent of GDP in 2008; the corresponding figure for Korea is 26.5 percent, the United States 26.1 percent, and the OECD average 34.8 percent (OECD, 2011b). If we exclude oil-related revenue, the ratio drops by more than one-third. Mexico's tax revenues (excluding PEMEX royalties) as a percentage of GDP lag behind those in Argentina, Chile, and Brazil (OECD, 2011b).

It should be noted, however, that such ratios are not fully comparable across countries because countries use their tax and transfer systems differently. For instance, Mexico uses tax credits instead of social transfers to compensate low-income wage earners in the formal sector, implying (all else being equal) a somewhat lower level of taxes or spending than in countries where transfers are used to compensate individuals. However, even if tax credits are taken into account (6.3 percent of GDP in 2003), including tax incentives for promoting investment and research and development (R&D) activities, Mexico still ranks lower than other OECD countries.

Income Tax and Value-Added Tax

Income tax and VAT have been important sources of revenue for Mexico, accounting for a combined 45 percent of revenue in 2007 (OECD, 2007c). In 1985, their combined share was only about 22 percent of all revenue. But, as Table 9.3 indicates, Mexico has a relatively low reliance on personal and corporate income taxes (28.6 percent of revenue, in comparison with 33.1 percent for the OECD average and 40.5 percent for the United States). To balance this, Mexico has a particularly high reliance on taxes on goods and services—50.2 percent of total revenue, the highest in the OECD—compared with 32.5 percent for the OECD average and 18.5 percent for the United States. For VAT and general sales taxes alone, we find that the ratio for Mexico is 19.7 percent, about the same as the OECD average but still well above the U.S. figure of 8 percent (not shown) (OECD, 2011b).

Oil Taxes

Around one-third of federal government revenues in Mexico stem from oil taxes, mainly in terms of royalty payments from the state-owned oil company, PEMEX (SHCP, undated). The risks of this dependence

Table 9.3
Mexico's Tax Revenue Sources Relative to Those of Selected Countries in the Organisation for Economic Co-Operation and Development, 2009

		Percentage of Total Tax Revenues				
				Social Security Contributions		
Country	Tax/GDP	Income Tax	Corporate Income Tax	Employees	Employers	Taxes on Goods and Services
United States	24.1	33.6	6.9	12.1	13.6	18.5
Germany	36.3	25.3	3.6	17.0	18.2	29.7
United Kingdom	35.0	30.5	8.1	7.8	11.4	29.0
Mexico	17.4	28.6		16.7		50.2
Korea	25.5	14.2	14.4	9.4	10.1	32.0
OECD average	33.8	24.7	8.4	9.4	15.1	32.5

SOURCE: OECD, 2011b.

NOTE: Mexico reports a combined share of personal income tax and corporate income tax, as well as social security contributions by employees and employers.

became evident in 1998, when oil prices fell by almost one-third. In 2000, Mexico created the Mexico Oil Revenue Stabilization Fund using the record windfall realized that year, with the intention of smoothing the impact that abrupt movements in world oil prices would have on public finances. Typically, less than half of revenue windfalls go into the fund. If oil export revenues are lower than projected in a given year, public spending could be financed with disbursements from the fund, limited to no more than half the fund balance at the end of the previous year and subject to the price of oil falling by a specified amount. If the revenue shortfall exceeds that amount, expenditure cuts must be made to complement resources withdrawn from the fund.[1]

Income Taxes

For statutory tax rates, we find that, in 2010, Mexico's top marginal rate of personal income tax (30 percent) was lower than those of the United States (35 percent) and the average for OECD countries (35.3 percent); many OECD countries have top rates of more than 40 percent. As shown in Table 9.4, Mexico's corporate income tax rate of 30 percent in 2011 was at the middle of the range of those for OECD countries, which varied from 12.5 percent (Ireland) to 39 percent (Japan and the United States), but significantly higher than in some Latin American countries, such as Chile. The corporate income tax rate was set at 28 percent on January 1, 2007, with no further reductions planned in the coming years. However, in 2009, Mexico changed its tax laws so that the corporate tax rate would be 30 percent for 2010–2012, 29 percent in 2013, and once again 28 percent in 2014 and future years. Also, the maximum individual income tax rate was set at 30 percent (Alvarado Nieto and Carbajo, undated; North American Production Sharing, undated).

---

[1] Mexico has started hedging against the volatility of oil prices by purchasing put options. For instance, Mexico bought 330 million barrels' worth of put options for 2009, to secure a minimum price of $70 for its crude oil ("Mexico Hedges to Protect Oil Revenues," 2008).

Table 9.4
Corporate Income Tax Rates in Selected Countries, 2000, 2006, and 2011 (%)

Country	2000	2006	2011
Japan	40.9	39.5	39.5
Germany	52.0	38.9	30.2
Ireland	24.0	12.5	12.5
Mexico	35.0	29.0	30.0
United States	40.0	40.0	39.2
OECD average	33.6	28.7	25.5
Chile	15.0	17.0	20.0

SOURCES: Chamberlain, 2006; OECD, undated (a).

NOTE: The income tax rates shown are the combined basic (nontargeted) central and subcentral maximum marginal tax rates.

Tax Policy Challenges

The low statutory rates partly account for the low tax-to-GDP ratio. Numerous tax exemptions and special regimes, a high level of informality, and weakness in tax administration also account for the low revenues. For instance, Mexico has the narrowest VAT base in the OECD, with tax paid on only 30 percent of consumption expenditures. This contrasts with figures of more than 50 percent for most OECD countries and an OECD high of 96 percent for New Zealand. Also, companies in certain industries, depending on the size of the individual company, can qualify for corporate income tax reductions. Those engaged exclusively in agriculture, livestock, forestry, and fishing enjoy a 50-percent reduction.

These preferential regimes, besides diluting revenue and complicating administration, create significant loopholes, distort economic activity, and facilitate evasion.[2] They also create a perception of unfairness, which reduces the willingness to pay taxes. A rough calculation suggests that the direct effect of completely eliminating the major preferential regimes could provide several percentage points of GDP in additional tax revenues (Cotis, 2003).

A large informal sector, which avoids paying taxes, is another reason for Mexico's low tax collection. Studies by INEGI show that informal activity accounts for about 20 percent of the profits generated by the economy. A study by the International Monetary Fund (IMF) (2006a) found that the informal sector represents more than 50 percent of the economically active population (EAP). About 30 percent of nonagricultural jobs are in the informal sector, with the service sector employing the greatest number of workers.

Tax Policy Reform

A marked increase in nonoil tax revenue cannot be expected without significant fiscal reform. In September 2007, Mexico enacted new tax reform to improve corporate tax collection and take a first step toward eventually eliminating the more complex traditional corporate income tax system. The bill proposes an alternative flat tax on business, known as the Flat Rate Business Contribution (Contribución Empresarial de Tasa Única, or CETU). Starting at a rate of 16.5 percent in 2008, the CETU rose to 17.0 percent in 2009 and to 17.5 percent in 2010. In practice, the CETU would act as a minimum corporate tax and as an alternative to the corporate income tax (that is, companies would have to pay either 28 percent minus all eligible deductions, or the flat tax, whichever is higher). The CETU eliminated the existing 2-percent corporate asset tax. One estimate (Economist Intelligence Unit [EIU], 2008) indicates that the new tax system would increase federal revenues by around 2 percent of GDP by the end of 2012.

Although these are important steps in the right direction, Mexico should further broaden the tax base and strengthen its tax administra-

---

[2] Mexico, like the United States, is a democracy, and one can place Mexico's difficulty in enacting tax reform in perspective by observing how contentious an issue it is in the U.S. Congress.

tion. Preferential regimes and deficiencies in tax administration remain the main concerns. Government officials will be hard pressed to break Mexico's informal economy from its culture of tax evasion.

Labor Market and Regulation

Labor legislation in Mexico is very detailed, complicated, and outdated. Mexico's labor regulation is contained in the federal labor law, which went into force May 1, 1970, and in Article 123 of the country's constitution, which dates back to 1917. This legislation regulates labor contracts, minimum wages, hours of work, legal holidays, and paid vacations, among other working conditions, as well as trade unions, strikes, and dismissal compensation.

As Figure 9.1 shows, labor regulations in Mexico have been among the most rigid in the OECD and emerging markets. They generally establish that the working relationship between an employer and an employee is permanent; severance payments are high. Restrictive

Figure 9.1
International Comparison of Labor-Market Rigidity

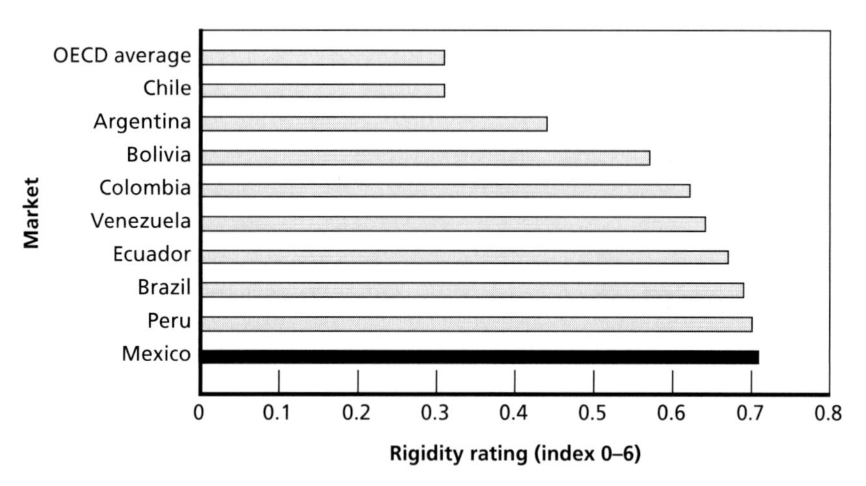

SOURCE: Botero et al., 2003.
RAND *MG985/1-9.1*

hiring and firing modalities could increase the cost of labor considerably and deter entrepreneurship by keeping firms informal.

Informal Economy

Informal workers do not receive job benefits, such as medical insurance and retirement plans. Trade unions and social security protection are rarely found in the informal sector. Furthermore, in many large-scale formal-sector enterprises, workers are often hired "off the books." According to the IMF (2006b), around 50 to 60 percent of the Mexican working population might be classified as informal.

Incentives to move into the formal sector are weak. According to INEGI data, between July 2010 and July 2011, a total of 853,778 jobs were created (nearly 60 percent of them in services), and 142,485 of them were in one-person firms. These jobs tend to be created in low-productivity and low-income areas. More than 70 percent of all firms have fewer than ten employees (Antón, Hernández, and Levy, 2011). In firms with more than 51 employees, in which productivity and wages are typically higher than in other firms, 318,159 jobs were created (INEGI, undated [b]). One analysis of Mexican data found that, from 2001 to 2010, for every 100 Mexicans who joined the working-age population on an annual basis, only 26 found formal employment (Hussein, 2011; INEGI, undated [a]). We therefore see significant polarization in job creation in Mexico and creation of jobs in businesses with ten employees or more as a high priority.

Unionization

Mexican labor is highly unionized. According to Deloitte Touche Tohmatsu (2005), nearly 40 percent of Mexico's workforce is unionized; only 12.1 percent of the U.S. workforce is unionized. Unions in Mexico represent some 80 percent of industrial workers in establishments with more than 20 employees. The Confederation of Mexican Workers, which claims to have some 6 million members, is the country's largest labor organization. Most of these workers belong to one of nine national labor federations. Only about 20 percent of unionized workers belong to single-company unions; the remainder belong to nationwide organizations. Companies sometimes deal with considerable jurisdictional strife among the unions because each union seeks to secure improved working conditions and higher salaries for its members (Deloitte Touche Tohmatsu, 2005).

Despite the high degree of unionization, industrial reorganizations, privatization, and downsizing during the 1980s resulted in massive layoffs and numerous labor concessions from unions to management regarding work practices. Unions were further weakened during the 1990s, when authority (including financial) over some important sectors (e.g., education) was transferred from the central government to the states. Now unions have to negotiate separate contracts with each state government rather than negotiating a single contract nationwide.

Wages

For a country that has strong employment protection legislation, Mexico's minimum wage is low. *OECD Employment Outlook* (OECD, 2006c) reports that the labor costs for minimum-wage workers relative to median-wage workers were 19 percent in Mexico in 2004, compared with 31 percent for the United States and 42 percent for the OECD as a whole.[3] According to Deloitte Touche Tohmatsu (2005), most workers in Mexico earn two to three times the minimum wage set by the National Minimum Wage Commission (Comisión Nacional de Salarios Mínimos, or CONASAMI). This leaves the unemployed (especially the young workers) with little incentive to leave welfare because working at the minimum wage is likely to net them the same as welfare benefits.

Energy Policy

As mentioned earlier, oil revenues are a major source of government revenues. Table 9.5 shows total government revenues, oil revenues, and their share in government revenues since 1990. The volatility in oil prices is reflected in oil revenues.

---

[3] The figure for Mexico was the same in 2005, the year of latest available data as of November 2001.

Table 9.5
Mexican Government Oil Revenues, 1990–2010

Year	Total Government Revenues (a) (millions of dollars)	Oil Revenues (b) (millions of dollars)	(b)/(a) (%)
1990	63,564.79	19,251.96	30.29
1992	95,072.77	23,252.26	24.46
1994	83,646.64	21,679.61	25.92
1996	73,726.59	26,229.18	35.58
1998	79,002.19	23,513.96	29.76
2000	125,764.15	41,628.77	33.10
2002	136,027.49	40,206.86	29.56
2004	158,011.97	56,856.40	35.98
2006	208,338.94	79,270.98	38.05
2008	213,939.42	78,864.70	36.86
2010	238,710.13	78,458.62	32.87

SOURCE: BANXICO, undated.

Unfortunately, these revenues are part of the annual budget and spent by the government but not reinvested in exploration or scientific development in PEMEX, the state-owned monopoly.

Oil Production and Reserves

In 2010, Mexico was the seventh-largest oil producer in the world and the third largest in the Western Hemisphere, behind the United States and Canada (Energy Information Administration [EIA], 2011). Its proven reserves as of January 1, 2011, amounted to 10.1 billion barrels of crude oil equivalent, 1.2 billion barrels of gas liquids, and 2.4 billion barrels of crude oil equivalent of natural gas (PEMEX, 2011a). Another source, BP's *Statistical Review of World Energy* (2011), put crude oil reserves at the end of 2010 at 11.4 billion barrels, down from 47.8 billion in 1997. Part of this significant drop was due to the initial overstatement by Mexico of its reserves. In 1997, Mexico dropped its reserves from 47.8 billion barrels to 21.6 billion when PEMEX was forced to comply with strict U.S. Securities and Exchange Commission rules. Proven reserves are the only type the U.S. Securities and Exchange Commission allows oil companies to report to investors.

As seen in Figure 9.2, Mexican oil production has dropped since 2004. Mexico produced about 3.0 million barrels per day (bpd) of oil in 2010, down from a peak of 3.8 million bpd in 2004 (BP, 2011). Also in 2010, Mexico was the United States' second-largest oil supplier; global exports of crude oil were 1.3 million bpd, with about 1.1 million bpd going to the United States and an additional 140,000 bpd of refined products going to the United States (EIA, 2011). Production was expected to average 2.85 million bpd in 2011 and 2.83 million bpd in 2012 (EIA, 2011). Exports have been falling as well.

At prevailing production rates, reserves in Mexico are expected to last between ten and 11 years. Analysts believe that Mexican oil production has peaked and that the country's production will continue to decline in the coming years. In 2010, the two main oil fields of Cantarell, which started operations in 1979, and Ku-Maloob-Zaap (KMZ)

Figure 9.2
Mexico's Oil Production, 1965–2010

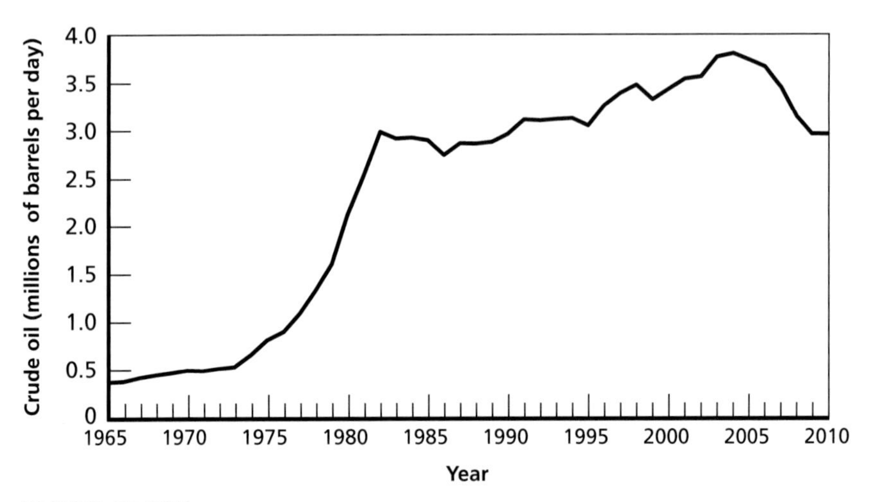

SOURCE: BP, 2011.
RAND *MG985/1-9.2*

were responsible for 54 percent of Mexico's crude oil production (EIA, 2011). In 2004, at its peak, Cantarell alone produced 2.12 million bpd of crude oil, or 62 percent of Mexico's total production. But, by 2010, it produced only 558,000 bpd. Cantarell production is expected to continue to decline (EIA, 2011).

The oil sector is an important component of Mexico's economy. Mexico's total energy consumption in 2008 consisted mostly of oil (58 percent), followed by natural gas (30 percent). Although the relative importance of oil to the general Mexican economy has declined, the oil sector still generates more than 14 percent of the country's export earnings (down from 68 percent in 1982) (EIA, 2011; "Mexico's Declining Oil Industry," 2007). Therefore, any further significant decline in oil production can have a direct effect not just on the country's overall fiscal balance but also financially, socially, and politically.

Challenges for Petróleos Mexicanos

A closer examination of PEMEX highlights many of the problems. Despite the high oil prices during most years since 2000, it registered a net loss of US$1.48 billion in 2007 ("Mexicans Divided on Proposed Energy Reform," 2008), whereas other major oil companies ended 2007 with strong profits. PEMEX continued to have losses in each year from 2008 to 2010, ending 2010 with a net loss of $3.8 billion (PEMEX, 2011b). According to Baker and Associates (2006), PEMEX has been the most heavily indebted oil company in the world. Transfers from PEMEX to the federal government have exceeded profits, which is one of the reasons for the company's heavy indebtedness. This debt, along with pension obligations and other liabilities, has hindered further access by PEMEX to international capital markets and prevented increased spending on exploration and production. Furthermore, the Mexican Congress must approve PEMEX's budget each year, which constrains PEMEX's ability to make independent funding decisions and hinders long-term planning.

The main challenges are sustaining the level of oil production and quality of oil by investing in exploration and development of oil-extraction methods (in particular, in reaching hard-to-access fields) in Mexico. To reverse this course, PEMEX could pursue exploration for new reserves in the deep waters of the Gulf of Mexico, but the company lacks the financial resources and expertise to do so. The World Bank (2006a) estimates that US$130 billion is required through the end of 2012 to restore the reserve rate and minimize imports.

In 2005, PEMEX reported possible new reserves of up to 54 billion barrels in the Gulf of Mexico. The company estimated that, if exploited, it could boost its total output to 6 million to 7 million barrels of oil per day ("Mexico's Oil Dilemmas," 2005). Foreign partners might be able to provide the technology and financial resources to explore these new fields. But Mexican law instituted as part of the oil nationalization of 1938 prohibits PEMEX from entering into profit-sharing ventures with other companies. Considering the significant long-term planning and development the oil industry requires, Mexico might be running out of time.

Despite its status as one of the world's largest crude oil exporters, Mexico is a net importer of refined petroleum products. Mexico's refinery capacity has not expanded at the same pace as the economy and the population. Domestic production of natural gas is similarly unable to meet demand, especially as natural gas is increasingly replacing oil as a feedstock in power generation. Mexico's oil production capacity, even if increased, would be insufficient to meet demand due to a significant shortage in refining and pipeline capacity. At one point, Mexico was estimated to be spending some US$14 billion yearly on imported gasoline, natural gas, and petrochemicals due to shortfalls in domestic production (J. Adams, 2006). Mexico has, however, started taking steps to rectify this. In March 2011, PEMEX began construction of a new 300,000 bpd refinery, the first new refinery built in Mexico in 30 years (EIA, 2011).

Oil Policy Reforms

Though much remains to be done, Mexico has made some progress toward addressing the challenges of the energy sector. In September 2007, the Mexican Congress passed a new tax reform bill to cut a general royalty on extracted hydrocarbons from MXN$0.79 on every peso of extracted oil to MXN$0.715 by 2012 and lighten the taxes PEMEX pays on oil drilled at mature or abandoned fields. It is estimated that this measure has taken $2.7 billion off PEMEX's 2008 tax bill and could amount to a savings of $5 billion in 2012 (Bremer, 2007).

In April 2008, President Felipe Calderón introduced a proposal to open the struggling oil monopoly to foreign investors. Calderón's proposal called for drilling in the deep waters of the Gulf of Mexico, something that PEMEX did not have the technology to do on its own. The proposal initiated a debate within Mexico on the future of its energy sector and PEMEX. One of the issues was whether this initiative violated the Mexican Constitution. The opposition denounced Calderón's plan as a move to privatize the nation's oil wealth, but Calderón and top officials at PEMEX insisted that, without radical reform, Mexico would run out of oil in less than a decade.

After more than five months of public debate, a compromise measure was passed on October 28, 2008. The energy reform legislation that Mexican Congress passed in October 2008 gives PEMEX more budgetary autonomy and transparency. The law also allowed some private investment through contracts that included risk-sharing but did not allow foreign companies to have a portion of total oil production or the price per barrel produced (U.S. Department of State, 2010b). In addition, some observers viewed implementation as slow and even largely unsuccessful (Camarena, 2010). Therefore, it does not appear that these reforms will do much to address declining production.

Fiscal Federalism

Table 9.6 lists the types of taxes collected by the various levels of the Mexican government and the taxes that are the shared responsibility of states and the federal government. Despite a federal governance structure, the federal government collects more than 90 percent of all revenues in Mexico (Diaz-Cayeros, 2004). State taxes vary but are generally much lower than federal taxes. Municipal taxes are even lower.

The Development of the Tax System

The fiscal arrangements in Mexico that are still in place in 2011 were established in the 1980s. With the aim of improving the effectiveness

Table 9.6
Federal and Local Government Taxes and Expenditures in Mexico

Responsibility	Tax
Federal	Corporate income tax Personal income tax VAT Duty on oil extraction (royalties) Oil export tax Tax on production and services (excises) Tax on new cars Tax on the ownership or use of vehicles Import duties Miscellany
Shared	Income taxes VAT Excises Tax on the ownership or use of vehicles Tax on new cars
State	State payroll tax Real estate transfer tax Tax on motor vehicles older than ten years Tax on the use of land Education tax Indirect taxes on industry and commerce Fees and licenses for some public services
Municipal	Local property tax Real estate transfer tax Water fees Other local fees and licenses Indirect taxes on agriculture, industry, and commerce Residential development

SOURCE: Sobarzo, 2004.

and efficiency of the tax system, states were asked to voluntarily cede the application of some indirect taxes, thereby avoiding the problem of double taxation. Shah (2004) argues that, in Mexico, the dominance of the central government resulted from both the direct assignment of functions to the federal level and the supposed inability of lower governments to assume delegated responsibilities. In return, the states started receiving revenue-sharing transfers from the federal government, the distribution of which was determined through a formula based on the states' population size and their relative fiscal efforts. As a result, the federal government was entitled to levy the most-important taxes in the country, such as the income tax, the VAT, and special taxes on production and selected services. These three taxes together represent nearly 90 percent of all federal tax revenues (Moreno, 2003).

Reforms passed in 2001 enabled individual states to impose taxes, not exceeding 5 percent, on companies or professionals with income of less than MXN$4 million. Since January 1, 2002, states have also been allowed to collect a sales tax of 3 percent and an income tax of 35 percent. However, few states exercise this right, given the potential political backlash. Most states do not collect sales or income taxes of their own, and state and municipal levies are largely limited to property taxes, service assessments, fees, permits, and some transport tolls (Moreno, 2003; Sobarzo, 2004).

Moreno (2005) explains that the lack of political competition characterizing the 1929–1988 period in Mexico, and the extraordinary influence of the Mexican president over the political careers of any elected post in the country, could have contributed to the subnational governments ceding their taxing authority to the federal government. During its 70-year hegemony, the Institutional Revolutionary Party (Partido Revolucionario Institucional, or PRI) maintained monopoly control of all levels of government, leaving the governors without incentives to question centralized tax collection because doing so would have meant the end of their political careers. Now that electoral competition and party alternation are in place in Mexican politics at all levels, the system of fiscal intergovernmental relations requires major changes, such as allowing subnational governments "to recover their taxing authority in order to fulfill the demands of local constituencies" (Moreno, 2005, p. 11).

However, according to Diaz-Cayeros, González, and Rojas (2002), the central problem of fiscal federalism in Mexico is that subnational governments do not have clear jurisdictions over virtually *any* policy area because the federal government always holds a crucial piece of the process. They further argue that, if political institutions remain as they are, greater decentralization will be counterproductive: Rather than producing a convergence between citizen demands and resource allocation, it will more likely widen the gap between who collects and pays taxes, reducing accountability and increasing regional tensions. Given the absence of consecutive reelection of legislators and incentives for reelection, Diaz-Cayeros, González, and Rojas (2002) note that local politicians are less likely to be accountable to the electorate than state or national politicians are because their careers depend more on their loyalty to party leaders.

Fiscal Transfers

The federation provides general (or unconditional) transfers, specific (or conditional) transfers to subnational governments, and revenue from federal taxes (so-called shared taxes), which are collected by the states. Although transfers from the federal government represented 8.1 percent of GDP in 2006, they were 6.8 percent as recently as 1998 (Ahmad et al., 2007). States in Mexico get almost 90 percent of their total revenues from federal transfers. In contrast, prior to the 1980s reforms, in the 1970s, states raised approximately 60 percent of their own revenue (Weingast, 2003).

Initially, the formula to calculate transfers to the states was based on need, usually proxied by population, and fiscal effort, derived as "the ratio of actual taxes collected to potential taxes estimated on the basis of some standard measure of fiscal capacity and some standard (e.g., national average) tax rate" (Bird and Smart, 2001, p. 19). The transfer system, however, does not provide much incentive for tax effort but treats smaller states (for example, Baja California Sur, Campeche, and Colima) favorably. Therefore, these states receive more per capita transfers than the others.

Using data in Hanson (2006), we find that the correlation between total population of the states and per capita transfers received by the states is negative (–0.61). These transfers are also non-redistributive. The correlation between per capita income of a state and the transfers it receives is weakly positive. Interestingly, the correlation between the number of people from a Mexican state who immigrated to the United States and that state's per capita transfers from the federal government is negative (–0.56).

Fiscal transfers, which are based on fixed formulas, do not appear to provide "insurance" to states—that is, provide relief to states that are economically worse off during a particular year, say, to finance unemployment benefits. Federal and state governments have "revenue stabilization funds," which can be used when revenues are below forecasts. However, Burnside and Meshcheryakova (2005) examine Mexican fiscal data from 1980 to 2003 and find that transfers to states, as well as other transfers, such as aid and social assistance, are highly procyclical (high when GDP is high) instead of countercyclical.

Individual State and Federal Financial Support

There are huge disparities across states in the amount of taxes collected. For example, in 1999, local governments in Oaxaca collected, on average, $55 per capita in own tax revenues (that is, tax revenue that is collected by the state itself rather than shared state and federal taxes); local governments in Nuevo Leon collected more than $425. Clearly, these differences are explained to a great extent by the size of the economy in each region, but regional economic wealth does not account for the whole variation. Quintana Roo and Campeche, for example, have equivalent economic and population sizes, but the latter generated 1.4 times more in own tax revenues than the former. Local governments of Queretaro and Guerrero collected roughly similar revenues per capita, even though the GDP per capita in the former is more than double the latter.

The most-important tax revenue sources for the states are the tax on older motor vehicles, a state payroll tax, and some indirect taxes on industry and commerce. In addition to these taxes, states can also charge fees for some public services (shows, entertainment, and urban public transportation). States have to pay a statutory general transfer of at least 20 percent to their municipalities out of the transfers they receive and all shared taxes. States are mandated to determine the formula for the distribution of the proceeds to their municipalities. These formulae have different aims and structure but, in many cases, are not transparent or published (Moreno, 2003; Sobarzo, 2004).

Regional Governments and Federal Financial Support

Regional governments in Mexico are as heavily dependent as state governments are on federal aid. According to Moreno (2003), unconditional transfers in 1999 represented 53.1 percent of total municipal revenues, followed by conditional or earmarked transfers, which accounted for 19.3 percent of revenues. Own-source taxes—the local property tax being the most important—represented only 10.4 percent of total revenues. This share is considerably below international standards. In the United States, for example, property taxes represented 73 percent of own tax revenue in 2005 and 25 percent of all local revenue, including state and federal transfers, in the same year (Prante, 2006).

The relative weight of subnational government expenditures is somewhat greater than that of taxes (see Table 9.7). According to the IMF's *Government Finance Statistics Yearbook, 2006* (IMF, 2007b), about 32 percent of the general government outlays in 2000 were carried out by state and municipal governments. In contrast, other federal countries ranked higher in this indicator in the same year, such as Brazil (42.8 percent), Argentina (41.3 percent), and the United States (48.6 percent) (Moreno, 2005). The bulk of state government expenditure is concentrated on public order and safety, transportation, health, education, and sanitation. Spending responsibilities of municipal governments include the provision of basic local infrastructure, such as water supply, sanitization, public safety, and local transportation.

Table 9.7
Structure of Revenues and Expenditures, by Level of Mexican Government, 2000

Revenue or Expenditure	Central Government (%)	State Government (%)	Local Government (%)
Total revenue	63.4	30.6	6.1
Taxes	50.1	12.6	3.6
Social contributions	6.6		
Grants		15.6	1.7
Other revenue	6.6	2.4	0.9
Total expenditure	68.2	27.5	4.3

SOURCE: IMF, 2007b.

Decentralization

The ongoing decentralization process (such as the step mentioned earlier to allow states to collect their own taxes) in Mexico could have significant implications for fiscal management and macroeconomic stability. A fuller decentralization would require reforms of the government structure and a substantial strengthening of public financial management, especially at the state and municipal levels. For instance, the federal government could consider giving the states and municipalities more taxing powers.

One of Mexico's greatest development challenges is the coexistence of extreme poverty in some regions with the economic dynamism of richer areas. The task for the central government is to identify a federal arrangement that can mitigate this disequilibrium using its fiscal mechanisms, so as not to provoke potential social and political conflicts or future instability.

U.S.-Mexican Economic Ties

The economic ties that bind the United States and Mexico have evolved over time. In this section, we examine how Mexico's opening up culminated in the formation of NAFTA, then turn our attention to NAFTA's effects on trade and investment.

Mexico's Reorientation to the World

Strong U.S.-Mexican economic relations are the result of policy choices by both countries and differ greatly from the economic relations that existed throughout most of the 20th century. Much of this reorientation came in the 1980s and 1990s as a response to economic crises and Mexico's attempt to restart economic growth (Lustig, 1998, 2001; Shatz and López-Calva, 2004).

As with many Latin American nations, Mexico followed a model of import-substituting industrialization throughout the 1950s and 1960s, with high trade barriers and strong government control of the economy. This paid off in high rates of growth and better living standards, but it also introduced economic distortions. Toward the end of the 1970s, Mexico started accumulating unsustainable levels of foreign debt. A debt crisis in 1982 ushered in years of economic stagnation and increasing poverty. It also brought in a rethinking of Mexico's orientation to the world economy.

Mexico's new strategy was formulated in 1984, with trade liberalization starting in 1985 and included removing trade and investment restrictions and diversifying exports in a unilateral liberalization strategy. In 1986, the country joined the General Agreement on Tariffs and Trade, the main international venue for multilateral trade liberalization and the predecessor organization to the World Trade Organization (WTO). In 1989, Mexico reformed its regulations regarding foreign direct investment (FDI), significantly broadening the right of majority foreign ownership, previously highly restricted; opening nearly all sectors to minority ownership; and expanding the use of an instrument known as the fee trust (*fideicomiso*) to allow foreigners to invest in industries limited to Mexicans. These investment liberalization rules were codified in a new investment law in 1993. As part of the reforms, Mexico allowed the free remittance of foreign investment–generated profits and the free repatriation of foreign-investment capital. These reforms made Mexico one of the fastest liberalizing countries in the world between 1986 and 1995 (Shatz, 2000).

The Economic Significance of the North American Free Trade Agreement

Mexico's economic relationships received their strongest boost on January 1, 1994, with NAFTA, a trade and investment agreement between Canada, Mexico, and the United States.[4] Even with the reforms of the 1980s, Mexico had still not started growing. In 1990, therefore, Mexico called for NAFTA as a means of increasing confidence in the reformed Mexican economy and restarting growth (Lustig, 1998). The agreement is wide-ranging in the subjects it covers. At the time of its

---

[4] For some, the move was not just of economic importance but of cultural importance as well, signaling that Mexico was a North American country, not a Latin American country (Rodriguez, 2005).

passage, NAFTA had the widest economic gulf between its members' living standards of all the reciprocal trade agreements in the world.

Mexico's economic relations with the United States dominate its economic relations with the rest of the world. However, Mexico has also tried to diversify its partners, not only through participation in the WTO but also through other multilateral forums and by aggressively seeking other bilateral and regional trade agreements. In 1994, the same year that NAFTA went into effect, Mexico joined the OECD, the multilateral grouping of mostly economically advanced nations. As of late 2010, Mexico had in force 12 free-trade agreements (FTAs) with 44 countries. Aside from the United States, the largest signatories to FTAs include the European Union (EU), in an FTA effective in 2000, and Japan, in an economic partnership agreement effective in 2005 (Secretaría de Economía, 2007b; NAFTA Office of Mexico in Canada, undated). As of 2011, Mexico also had 27 bilateral investment treaties in force, covering 28 countries, with one more signed and awaiting legislative approval (Secretaría de Economía, undated [a]).

The results of Mexico's opening have been dramatic. Between 1980 and 2010, total merchandise trade rose from $39.1 billion to $599.8 billion, and exports rose from $18.0 billion to $298.4 billion (BANXICO, undated). More important to Mexico's economic diversification is the fact that petroleum exports as a share of total merchandise exports fell from a peak of 68.5 percent in 1982 to a trough of only 6.2 percent in 1998. Some of this decrease is due to oil prices: Oil prices were historically high in 1982 and historically low in 1998. But even with the very high oil prices of 2008, petroleum constituted only 17.4 percent of Mexico's merchandise exports, lower than in every year from 1980 to 1992 (BANXICO, undated). Although Mexico trades around the world and absorbs investment from many countries, the United States is by far its most important partner.

U.S.-Mexican Economic Relations Since the North American Free Trade Agreement Went into Effect

The trade and investment relationship between the United States and Mexico is one of the largest in the world. From 2006 to 2010, Mexico was the third-largest trading partner for the United States in goods and services, behind only Canada and China, with total trade equal to $332.2 billion in 2006, $346.8 billion in 2007, $342.6 billion in 2008, $340.8 billion in 2009, and $431.2 billion in 2010. In 2005, it was the United States' second-largest trading partner, with trade in goods and services of $290.5 billion, behind only Canada (U.S. Census Bureau, 2006c, 2007c, 2008a, 2011; Bureau of Economic Analysis, 2008; Koncz-Bruner and Flatness, 2010, 2011).

In 2010, merchandise exports to Mexico constituted 12.8 percent of all U.S. merchandise exports, and merchandise imports from Mexico constituted 12 percent of all U.S. merchandise imports. This made Mexico the second-largest merchandise export destination and the third-largest merchandise export source. The numbers are much lower for FDI—cross-border investment for the purpose of controlling or operating a business. In 2010, U.S. FDI in Mexico amounted to only 2.3 percent of worldwide U.S. FDI, measured as direct investment position, or total stock of FDI. Mexican FDI in the United States amounted to only 0.5 percent of total FDI in the United States (Barefoot and Ibarra-Caton, 2011). Although the United States trades heavily with Mexico, the leading investment partners of the United States are European countries.

Mexico's trade and investment dependence on the United States is far higher. In 2010, a total of 79.9 percent of Mexican merchandise exports went to the United States, the lowest share since at least 1993, if not before (Secretaría de Economía, 2011a). In 2010, 48.1 percent of Mexico's merchandise imports came from the United States, a slightly higher proportion than in 2009 (48.0 percent), but the second lowest since NAFTA went into effect. The highest was 75.5 percent, reached in 1996 (Secretaría de Economía, 2011b). In terms of FDI, only slightly more than one-quarter of FDI flows to Mexico in 2010 came from the United States (27.6 percent), but that was an unusual year. That figure averaged 53.9 percent between 1994 and 2010 and peaked at 71.8 percent in both 2000 and 2001 (Secretaría de Economía, undated [b]).

One of the hallmarks of U.S.-Mexican trade is its high degree of intracommodity flows. This is because, in large part, trade between the United States and Mexico takes place as part of production-sharing arrangements, with inputs exported from the United States to Mexico, processed or finished in Mexico into final goods, and then sent back to the United States. The top three manufactured commodity import and export groups have consistently been electrical machinery and equipment, nonelectrical machinery, and motor vehicles and parts. A more detailed analysis of trade for 1989 to 2002 showed that Mexico was always among the top four U.S. trading partners in terms of the level of intracommodity trade. During that period, an index of intracommodity trade for the United States with the world averaged 47.9 percent, whereas the index with Mexico averaged 67.5, with 0 signifying no intracommodity trade and 100 signifying total intracommodity trade (Shatz and López-Calva, 2004).[5] Some economies, such as those of Hong Kong, Taiwan, and Korea, have achieved growth by starting with production-sharing arrangements and then gradually moving into original-equipment manufacturing and own-brand production. In fact, Carlos Salinas, when he was president, said that Korea was a model for Mexico (Hanson, 2010). However, unlike these other countries, Mexico has not moved beyond production sharing as its main form of manufacture for export. Indeed, as of 2006, Mexico's production-sharing industries accounted for 20 percent of Mexico's manufacturing value added and almost half of Mexico's exports (Bergin, Feenstra, and Hanson, 2009). NAFTA might have even encouraged the growth of production sharing by allowing Mexico to shift domestic resources to export production, which has experienced higher volatility than the overall Mexican economy (Bergin, Feenstra, and Hanson, 2009).

The dominance of production sharing and assembly in the U.S.-Mexican economic relationship might present a special challenge to Mexico because of the rise of China as one of the premier locations for assembly in the global economy. In particular, Chinese manufacturing is a substitute for Mexican manufacturing (Kehoe and Ruhl, 2010). China outstripped Mexico in the share of total U.S. imports after 2001, the year China acceded to the WTO (Hanson, 2010). However, Mexico faces tougher competition from other countries in both the

---

[5] The index is the Grubel-Lloyd intraindustry trade index, explained in Grubel and Lloyd (1975), and computed at the two-digit level of the Harmonized Tariff Schedule codes.

world and U.S. manufacturing markets (Chiquiar and Ramos-Francia, 2009).[6]

According to our calculations using INEGI data, because of the level of production sharing, much of the FDI in Mexico is concentrated in the six states along the U.S. border, along with Mexico City, the major population and economic center in the country. Between 1999 and 2010, those seven entities attracted between 73 percent and 89 percent of FDI flows into Mexico each year. In contrast, during the period 1999 to 2009, they averaged 45 percent of Mexico's GDP and, according to the 2005 population census, had only 26.1 percent of the Mexican population.

Mexico's three southernmost states—Chiapas, Oaxaca, and Guerrero—attracted less than 0.4 percent of all FDI to Mexico between 1999 and 2006. In contrast, they produced, on average, 4.9 percent of Mexico's GDP and had 10.6 percent of its population (INEGI, 2006, undated [a]; Secretaría de Economía, 2010).

This discussion can be put into temporal perspective by examining the time trends for trade and investment since 1994. Merchandise trade between the United States and Mexico has grown strongly. Since at least the year in which NAFTA went into effect, merchandise trade between the United States and Mexico has grown faster than merchandise trade between the United States and the rest of the world (see Figure 9.3). Much of this speedier growth took place between 1994 and 2002; however, after three years of slower growth, U.S.-Mexican trade growth again outstripped U.S.–rest-of-world trade in 2006, again grew more slowly in 2007 and 2008, and then grew much more quickly in 2009 and 2010.[7] In fact, growth of trade with Mexico in 2009 and 2010 was about 7 percentage points higher than growth in trade between the United States and the rest of the world in both those years.

---

[6] In worldwide manufacturing markets, these competitors include Hungary, Korea, Philippines, Poland, Thailand, and Turkey. In the U.S. manufacturing market, these competitors include Hong Kong, Indonesia, Korea, Malaysia, Philippines, Taiwan, Thailand, and Turkey (Chiquiar and Ramos-Francia, 2009, p. 18, Figure 6).

[7] Pastor (2008) also notes the slowdown of U.S.-Mexican trade after 2000.

Figure 9.3
Growth of U.S.-Mexican and U.S.–Rest-of-World Merchandise Trade, 1994–2010

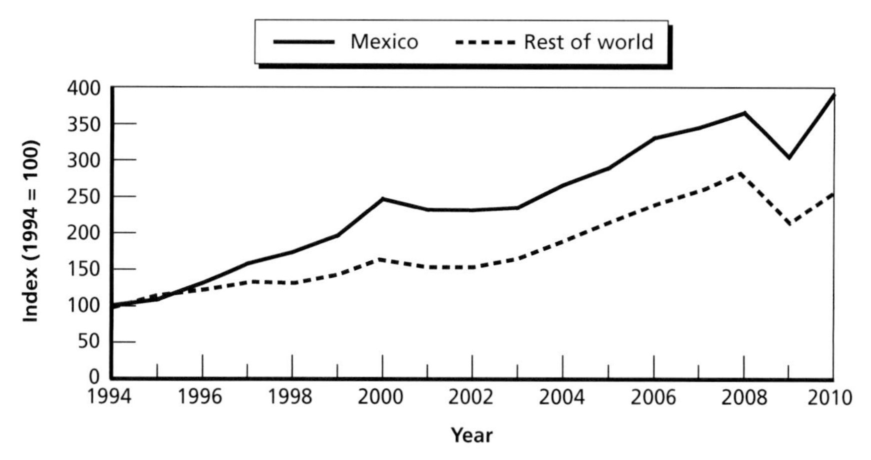

SOURCE: U.S. International Trade Commission, undated.
NOTE: Data are indexed to nominal values.
RAND *MG985/1-9.3*

Just as U.S.-Mexican trade has grown rapidly, U.S.-Mexican FDI has grown rapidly. In fact, it has grown even more rapidly than U.S.-Mexican trade (see Figure 9.4). Between 1994 and 2010, an index of total two-way U.S.-Mexican FDI, measured as the direct investment position, or stock of FDI, grew from 100 to 530, whereas an index of total FDI between the United States and the rest of the world grew from 100 to 517. However, after years of more-rapid growth of FDI with Mexico, the situation reversed in 2010, when the respective indexes were 541 and 572.

Some of this growth in trade and investment would have occurred regardless of whether NAFTA existed or not. As noted earlier, Mexico changed its own trade policies unilaterally and embarked on a series of internal reforms as well. In addition, Mexico had already received tariff preferences on some goods from the United States. However, NAFTA further enlarged those preferences and created a process for investment dispute resolution that was intended to make Mexico a more inviting site for FDI. Empirical research has found that NAFTA did contribute independently to the growth of trade between the two countries.

Figure 9.4
Growth of U.S.-Mexican and U.S.–Rest-of-World Foreign Direct Investment, 1994–2010

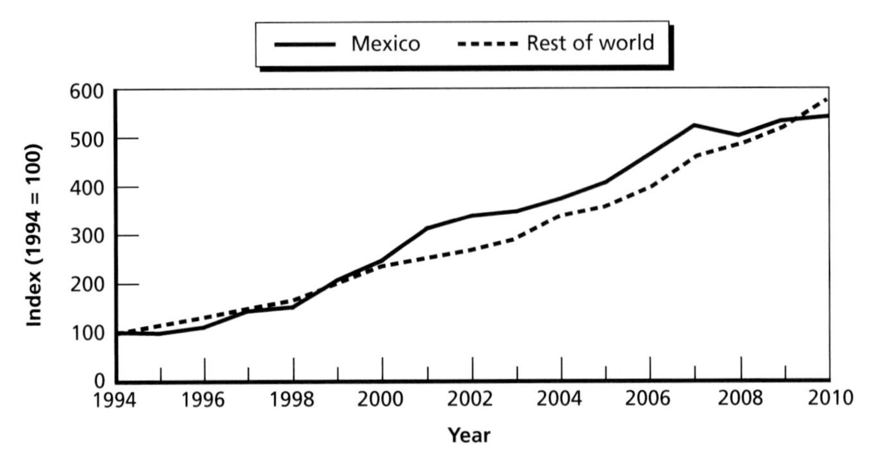

SOURCE: BEA, 2010.
NOTE: Data are indexed to nominal values.
RAND *MG985/1-9.4*

On average, a 1-percentage-point increase in the U.S. tariff preference for Mexican goods (stated another way, a 1-percentage-point decrease in tariffs faced by Mexico relative to tariffs faced by other countries) increased Mexican exports to the United States by 6.9 percent. This relationship was greater during the post-NAFTA period than in the period before NAFTA. In addition, a 1-percentage-point increase in the tariff preference offered to the United States by Mexico is associated with a 5.4-percent increase in U.S. exports of the specific good (McDaniel and Agama, 2003).

NAFTA has also changed patterns of trade for other countries. Not only did Mexico's share of U.S. imports grow most rapidly for those goods with the largest NAFTA tariff preferences, but Mexico's share of imports from the EU declined for those products that have large NAFTA preferences (Romalis, 2005).

The North American Free Trade Agreement Trucking Legislation as an Example of Bilateral Conflict Resolution

The success of NAFTA was tacitly based on unrestricted shipping between North American countries. Given the common land borders, trucking is clearly an important mode of shipping goods between these countries. Trucks from neighboring nations were to be given access to border states by 1995 and regionwide by 2000. The provisions have been in effect for truck movements between Canada and the United States; however, until July 2011, Mexican trucks had been excluded due to alleged safety and other concerns, despite NAFTA requirements (U.S. Department of Transportation, 2011).

That aspect of NAFTA was stalled by lawsuits and disagreements between the United States and Mexico. Organized labor has long objected to opening the United States to Mexican trucking, which labor leaders fear will depress wages and decrease the number of jobs in the United States. Safety groups were concerned about the maintenance and safety of Mexican fleets. In addition, increasing concerns about illegal immigrants from Mexico and threats of terrorist activity made more likely by porous borders, especially the southern one with Mexico, have increased the pressure to restrict the access of Mexican trucks into the United States. Conversely, business groups have wanted the border opened to avoid the costs of employing middlemen to transfer goods from Mexican to U.S. trucks. Mexican truckers are also more cost-competitive, earning only about one-third of their American counterparts.

The number of northbound cargo truck crossings through the U.S.-Mexican border increased 163 percent between 1994 and 2004, from 2.76 million to 4.5 million. Until full implementation of unrestricted cross-border trucking under an agreement reached early in 2011 (described later in this monograph), Mexican trucks are required to drop off United States–bound shipments within the immediate border zone, generally 25 miles from the international border. Likewise, U.S. trucks must drop off shipments in Mexico shortly after crossing the border. Most of the border crossings are actually done by so-called drayage truckers. Long-haul trucks, which drop off trailers (cargos) at forwarders on their side, transfer their loads to the drayage truckers, who transport these trailers across the border. Long-haul trucks on the other side then pick up the trailers and take them to their destination. The complicated arrangement is time-consuming (taking anywhere from four to 23 hours) (Economic Research Service [ERS], 2000) and expensive. Mexico estimates its losses at $2 billion annually; U.S. shippers say they have incurred similar costs (Collier, 2001).

President Calderón's March 2011 visit to the United States resulted in an announcement that Mexico and the United States had found a clear path to resolving the cross-border trucking issue. In July 2011, the two governments were able to forge an agreement that would allow Mexican trucks that satisfy safety and environmental standards access to the United States for a trial period of three years. As part of the agreement, Mexico agreed to lift retaliatory tariffs it had instituted two years previously (U.S. Department of Transportation, 2011).

The amicable solution of this dispute demonstrates that even contentious issues between the two countries can be resolved. Further issues, however, remain. For instance, both countries are working to eliminate barriers to remaining sensitive products. For Mexico, this means authorizing the import of U.S. beans and corn; for the United States, it means authorizing the import of Mexican sugar.[8]

---

[8] A chronology of events related to the trucking issue can be found in the Federal Register notice about the July agreement (U.S. Department of Transportation, 2011).

CHAPTER TEN

Inequality, Poverty, and Social Policy in Mexico

In this chapter, we analyze the challenges Mexico faces in the social sector. This is a topic worthy of U.S. concern because economic reforms that promote growth but fail to reduce poverty, inequality, and regional disparities will not succeed in achieving sustainable economic growth and the objectives of these reforms, such as potentially reducing migration. Therefore, programs and policies that improve the well-being of the population need to be put in place in parallel with other structural reforms.

Inequality and Poverty

As we have discussed, many of Mexico's economic policies have focused on stabilizing the economy and promoting growth. Poverty and inequality are two important features of Mexico's society and are highly correlated. We begin this chapter by focusing on regional and social inequality.

Inequality

Although poverty rates declined from 1994 to 2006 (see Figure 6.4 in Chapter Six), the question remains whether this shift has translated into lower inequality, not only in income but also in wealth and human capital.[1] Figure 10.1 shows values of the income Gini coefficient for

[1] However, some of the gains in poverty reduction were reversed in 2008 (see Figure 6.4 in Chapter Six).

Figure 10.1
Inequality in Mexico and Other Countries in the Organisation for Economic Co-Operation and Development, 1980s–2000s

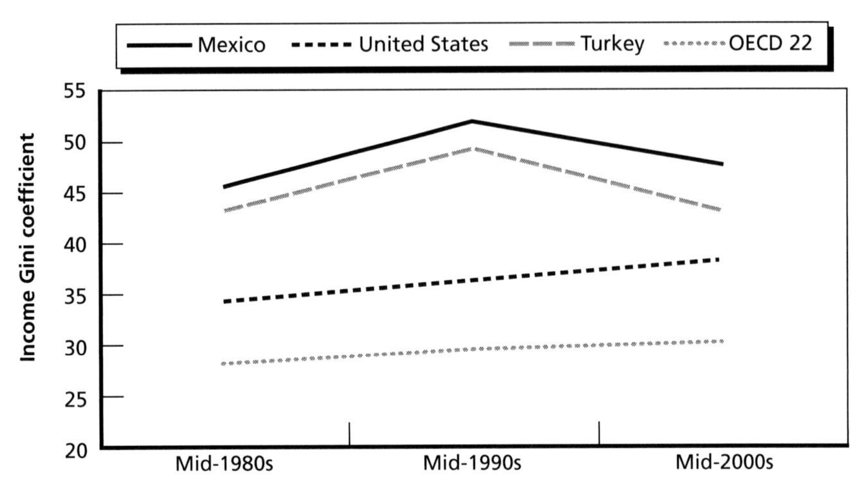

SOURCE: OECD, undated (b).
RAND *MG985/1-10.1*

Mexico, the United States, Turkey, and a group of 22 OECD countries at different points in time during the past three decades. The Gini coefficient is a measure of inequality; a value of 0 indicates "perfect equality," in which every person in an economy has exactly the same amount of, say, income, and a value of 100 indicates "perfect inequality," in which one person concentrates all income in the economy. As shown in the figure, income inequality in Mexico increased significantly from the mid-1980s and mid-1990s. Although inequality decreased from a Gini coefficient equal to 0.52 in the mid-1990s to 0.47 in the mid-2000s, it remains above the mid-1980s levels, and it is still significantly higher than in other OECD countries, including Turkey. The pattern of an initially increasing overall inequality from 1994 to 2000 and then a decrease in inequality until 2006 resulted in a combination of a steady decrease in inequality in urban areas during that period and an initial increase and then decrease in inequality in rural areas (Equivel, 2010). In 2008, inequality again rose (to a Gini coefficient of 0.52), as did poverty (World Bank, undated [c]).

A 2008 report indicates not only that Mexico ranked last in income equality among all OECD countries but also that Mexico's inequality is strictly greater than that for all OECD countries across all deciles in the income distribution. That is, Mexico's income distribution from top to bottom is worse than that of any other OECD country, unlike, for example, the income distribution of the United States, which is, at all income deciles, worse than that of any OECD country except Mexico (see OECD, 2008, Table 1.A2.1).

De la Torre and Moreno (2004) explores Mexico's distribution of income, wealth, and human capital between 1994 and 2002.[2] The authors find that inequality remained mostly constant during this period, which suggests that the period of economic expansion between 1996 and 2000 did not result in a significant decrease in wealth or human capital inequality. In fact, between 1998 and 2000, a period in which Mexico's GDP grew at an average of 5 percent per year, the wealth Gini coefficient grew from 0.52 to 0.57. As De la Torre and Moreno (2004) point out, this might indicate that the groups with the highest income were the ones that obtained the best opportunities to accumulate wealth during the phase of economic growth. It is worth noting, however, that De la Torre and Moreno's analyses do not cover the entire period shown in Figure 10.1, so it is entirely possible that, as income inequality decreased toward the mid-2000s, wealth inequality might have also decreased in the years subsequent to 2002.

Another feature revealed by De la Torre and Moreno is that human capital accumulation in Mexico is subject to less inequality than income and wealth, probably because the first nine years of school are freely provided by the Mexican government. Their analysis suggests that human capital inequality remained stable between 1994 and 2002. However, there were important changes in the distribution of educational achievement in that period that are not captured by changes in their model. Figure 10.2 reports the share of the population with

---

[2] De la Torre and Moreno's estimates of Mexico's income Gini coefficient differ somewhat from those computed by the OECD. This is probably due to the use of different definitions of income because the OECD uses disposable income (after taxes and transfers) to measure inequality.

Figure 10.2
Population and Income Shares, by Education Group, 1994 and 2002

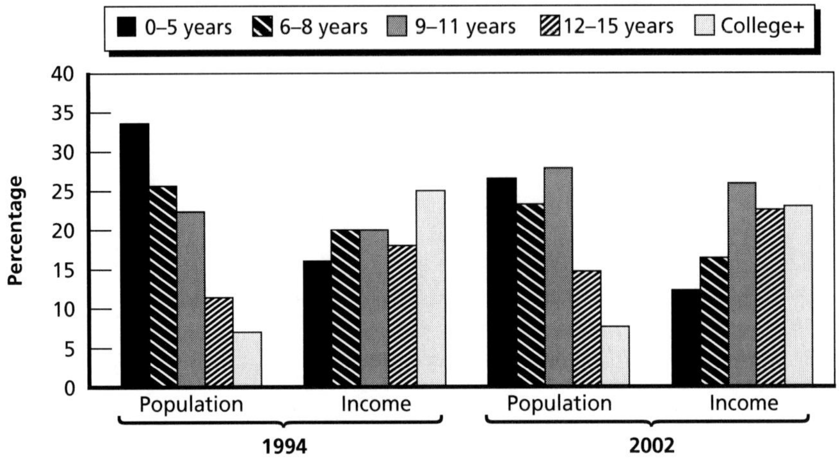

SOURCE: De la Torre and Moreno, 2004.
RAND *MG985/1-10.2*

different levels of education and the income share of those education groups in 1994 and 2002. A positive trend in educational achievement is suggested by this figure because the percentage of individuals in the two least-educated groups (zero to eight years of schooling) decreased from 59 to 50 percent of the population, while the percentage of individuals in the three most-educated groups (nine years of schooling or more) increased from 41 percent to 50 percent. By 2010, the percentage of individuals in the two least-educated groups had fallen further to 41 percent (INEGI, undated [a]).

On the other hand, the figure also shows that significant income inequalities still exist between education groups. Although the two least-educated groups constituted 50 percent of the population in 2002, they accounted only for 28 percent of total income, while the highest-educated group accounted for only 8 percent of the population but 23 percent of total income.

Regional Disparities

Regional disparities have been exacerbated in the past due to an increase in population dispersion. The number of rural communities increased by more than 100 percent from 1970 to 1995, while the total population in those communities increased by only 21 percent during the same period, which indicates that the increase in the number of rural communities was driven mostly by out-migration to urban areas.[3] INEGI defines a locality with a population of 2,500 or less as rural and those with populations larger than 2,500 as urban (INEGI, undated [a]).

In 2000, Mexico had 197,930 rural localities and 1,461 urban localities. Almost 30 percent of the population lived in the rural localities. The states that had the highest number of rural localities are Oaxaca, Chihuahua, Yucatan, Sonora, and Puebla. The dispersion of rural communities represents a challenge to the Mexican government because these areas are predominantly poor and it is difficult to provide them with infrastructure and services.

Rural Poverty

Most Mexicans living in rural areas are indigenous. The indigenous population in Mexico ranges between 12 percent and 30 percent of the population. The 30-percent estimate includes indigenous population that might have been assimilated into the Mexican mestizo culture, losing their original language and traditions (Instituto Nacional Indigenista, 2002). By 2005, the proportion of the indigenous population reported by the census was 15.01 percent (INEGI, 2010). CONAPO (2005b) projections estimated that the proportion of the total population that would be indigenous peoples would be 12.8 percent in 2010. There are 62 official indigenous languages, each with multiple dialects, and each is considered a national language with the same status as Spanish (Instituto Nacional Indigenista, 2002).

[3] Therefore, as people migrate out of medium-sized communities toward large urban areas, the total population of many previously urban communities falls below 2,500—and thus those communities become rural by definition—and the number of rural communities increases, but the overall sum of their population decreases.

The concentration of indigenous groups is mainly in the central and southeastern states. The states with the greatest percentage of indigenous population using data from 2000 are Yucatan (65.5 percent), Oaxaca (55.7 percent), Quintana Roo (45.6 percent), Chiapas (30.9 percent), Campeche (30.9 percent), Hidalgo (25.9 percent), Puebla (20.9 percent), Guerrero (18.6 percent), Veracruz (16.9 percent), and San Luis Potosí (16.8 percent) (CONAPO, 2005b).

Life Expectancy

According to the World Bank, the indigenous population faces extreme poverty in Mexico (Psacharopoulos and Patrinos, 1994; World Bank, 2004). The life expectancy for indigenous peoples in 2000 was 69.4 years for men and 74.7 for women; for the rest of the population, it was 71.5 and 76.5 years, respectively. By 2010, life expectancy for indigenous men was expected to be 72.5 and 77.7 for women, in comparison with the total population for men 74.2 and for women 79.0 (CONAPO, 2005b). The gap between the life expectancy of indigenous population and that of the total population has been very similar across time. The main causes of mortality for this population are contagious diseases: influenza and pneumonia. In 1995, child mortality among the indigenous population was 54 deaths per 1,000 live births, almost twice that of the rest of the population, but significant progress has been made since then: By 2006, that figure had been reduced to 27 deaths per 1,000 live births (Pan American Health Organization [PAHO], 1998, 2007). CONAPO (2005b) projected that, in 2010, child mortality would decline to 22.8 deaths per 1,000 live births in the indigenous population, in contrast to 14.0 deaths per 1,000 live births for the rest of the population.

Crop Production

The main crops in Mexico are corn, wheat, beans, other grains (such as rice), and sesame seeds. Mexico also produces cotton, barley, soybeans, sorghum, and safflower. Other crops include sugarcane, tomatoes, bananas, chilies, green peppers, oranges, lemons, limes, mangoes, avocados, other tropical fruits, blue agave, and coffee.

Rural Population and Agricultural Reform

The primary sector's contribution to total GDP declined from 19 percent in 1950 to less than 4 percent in 2009, as shown by the continuous line in Figure 10.3.[4] The decreasing economic importance of the primary sector has disproportionately affected rural areas. Despite the fact that those rural areas contain 25 percent of the country's population, they accounted for only 2 percent of GDP in 2000 (OECD, 2007b). It is not surprising, then, that living standards in rural areas are considerably lower than in the rest of the country (OECD, 2007a; Yúnez-Naude, Barceinas, and Ruiz, 2004).

In the second half of the 1980s, the Mexican government began a series of reforms that sought to liberalize the agricultural sector. These reforms, as well as others that aimed at increasing legal certainty in land ownership, culminated with the signing of NAFTA in 1992. The government hoped that competition and market forces would result in a shift in farming supply away from products in which Mexico was less

Figure 10.3
Rural Population Share of Total Population and Primary-Sector Share of Total Gross Domestic Product, 1950–2010

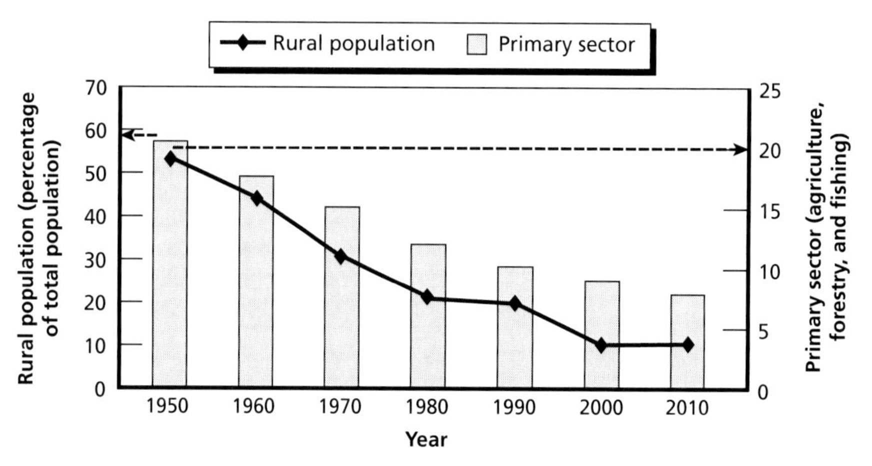

SOURCES: INEGI, undated (a); OECD, undated (c).
RAND *MG985/1-10.3*

[4] The primary sector is made up of agriculture, forestry, and fishing.

competitive (such as grains and oilseeds) and toward those that could be exported and in which Mexico was more competitive (such as fruits and vegetables). This would eventually lead to increased productivity and higher income for farmers, preparing them for the eventual effects of NAFTA: convergence of the prices of imports and exports with international prices, increase in the importance of agricultural trade, and a reduction in the production of importable goods, accompanied with an increase in the production of exportable goods (Yúnez-Naude, Barceinas, and Ruiz, 2004).

Some of these goals have indeed been achieved. By 2004, the value of Mexican tomato exports had doubled to US$1 billion per year, and exports of other fruits and vegetables increased substantially as well (Pérez, Schlesinger, and Wise, 2008). In addition, prices of several agricultural products—including corn and beans—have decreased considerably, allowing consumers to buy food at lower prices (OECD, 2006a). There is no consensus, however, regarding whether the overall effect of trade liberalization has been positive or negative. Pérez, Schlesinger, and Wise (2008) argue that NAFTA has (1) failed to generate adequate employment for those displaced from traditional agriculture, (2) not stimulated greater efficiency and productivity in Mexican agriculture, and (3) worsened Mexico's trade balance because import growth has outpaced export growth. A recent study of the effects of agricultural trade liberalization on poverty in 15 countries concludes that only two of them will experience an increase in poverty as a result of liberalization, one of them being Mexico (Hertel et al., 2007).

In addition, Yúnez-Naude and Barceinas Paredes (2004) conducted an econometric analysis of NAFTA's effect on Mexican agriculture. They find that the productivity and total land use in the production of most importables (such as corn and beans) did not experience major changes; on the other hand, they do find evidence that some exportables (tomato, broccoli, cantaloupe, and watermelon) experienced significant productivity increases as a result of NAFTA. They conclude that a trend of convergence of Mexican crop prices—both imports and exports—toward U.S. prices already existed before NAFTA and that the implementation of NAFTA might have accelerated convergence for exportable agricultural prices.

On the other hand, such institutions as the World Bank and the OECD continue to promote the liberalization of trade of agricultural products. The OECD (2006a) suggests needed agricultural reforms in Mexico, which include the elimination of market price support and a reform of the energy regime that facilitates the elimination of energy subsidies. The World Bank's 2008 World Development Report also points out market distortions as hampering greater efficiency in the agricultural sector (World Bank, 2007b). The report concludes that Latin American countries stand to benefit the most from further trade liberalization, as long as such liberalization is coupled with broader economic reforms that improve market infrastructure, institutions, and support services.

Social Policy

With that overview of inequality, regional disparities, and rural poverty, we proceed to a deeper analysis of the main social policies that could be influencing these patterns. We concentrate on education, health, health insurance and social security coverage, public social programs designed to alleviate poverty, and public programs designed to capture remittances from Mexican immigrants to the United States.

Education

Measured by the number of students, Mexico has the second-largest education system both in the OECD and in Latin America (behind only the United States and Brazil, respectively). Furthermore, the total number of students has steadily increased in the past three decades, more than doubling between 1970 and 2000, and growing by 9 percent just from 2000 to 2005. In 2002, there were more than 30 million students in the entire Mexican education system; almost 80 percent of them were enrolled in the basic education level, which covers preschool to lower secondary (Secretaría de Educación Pública [SEP], undated).[5]

---

[5] The education system is divided into four levels: Preschool provides education for children aged three to five; primary education includes grades 1 through 6; lower secondary

Certain educational levels have experienced faster growth than others, which is largely by explained the already-high enrollment rates in the primary and lower secondary levels, for which attendance is compulsory. Table 10.1 shows that the number of students in primary school remained essentially unchanged between 2000 and 2008, while both the number of students and enrollment rates experienced significant increases at the preschool and high school levels in the same period. Although the enrollment rates of both lower and upper secondary school have increased, there is still a significant number of youth who either never enroll in or drop out of upper secondary school. According to Santibañez, Vernez, and Razquin (2005), enrollment and dropout figures imply that, out of every 100 students who enter primary school, around 68 complete the first nine years of education and 35 graduate from upper secondary. Only 8 percent of Mexicans hold bachelor's degrees.

Because education in Mexico is provided primarily by the public sector, the increase in the demand for education services implied by the growth of the student population has placed significant budget pressures on the government.[6] For example, between 1995 and 2001, public spending on basic education grew by 36 percent, among the steepest increases in the OECD (Santibañez, Vernez, and Razquin, 2005). In 2003, public expenditures on education represented 5.6 percent of Mexico's GDP, higher than the OECD average of 5.2 percent. In 2007, Mexico's public expenditures on education as a percentage of the GDP declined to 4.8 percent (World Bank, undated [c]). Moreover, Mexico is the OECD country with the largest public expenditure on education as a share of total public expenditure—more than 20 percent, well above the OECD average of 13.3 percent (OECD, 2006b).

Although these expenditure figures correctly indicate that education is a high policy priority in Mexico, they disguise a very important

---

education covers grades 7 to 9; and upper secondary education covers grades 10 to 12. Both primary and lower secondary are compulsory, and preschool is in the process of becoming compulsory as well.

[6] The share of students enrolled in public schools is more than 90 percent in basic education, 80 percent in upper secondary, and about two-thirds in higher education (Guichard, 2005).

Table 10.1
Enrollment and Enrollment Rates in Mexico, by Education Level, 2000 and 2008–2009

	Students			Enrollment Rate (%)	
Education Level	2000 (thousands)	2008–2009 (thousands)	Change (%)	2000	2008–2009
Basic education					
Preschool	3,423	4,608	35	50	78
Primary	14,793	14,861	0	93	97
Lower secondary	5,349	6,128	15	82	95
Upper secondary					
Technical	362	373	3	—	—
High school	2,594	3,682	42	47	62
Higher education					
Teacher's college	201	104	−48	—	—
University	1,718	2,547	48	—	—
Graduate school	129	196	52	—	—
Total	29,621	33,876	14	—	—

SOURCES: SEP, undated; Robles Vásquez and Martínez Rizo, 2006.

NOTE: The enrollment rate is defined as the number of students enrolled in each level divided by the total population of theoretical age in that level (that is, the number of students who would normally be at that school level at that age).

issue: Almost 80 percent of public education expenditures go toward teacher compensation; when all education workers are considered, this figure is 91 percent (OECD, 2006b).[7] This is high not only by OECD

[7] These percentages are for primary, secondary, and postsecondary nontertiary (i.e., vocational training or educational programs for students who tend to be older than those enrolled at the upper secondary level) education. For tertiary education, teacher compensations account for 56 percent of total expenditures and compensation of all staff for 73 percent.

standards but also when compared with expenditures in other Latin American countries.

There is then little left to invest in other educational resources that might also be critical to improving educational quality. Given historical trends and the significant power and organization of the national teachers' union, any effort to change the structure of expenditure might prove to be a difficult one to undertake. Nevertheless, since 2001, Mexico started creating programs that focus on allocating resources to nonwage expenditures.[8] Furthermore, as Guichard (2005) discusses, some opportunities might arise in the medium and long runs: As the cohorts attending basic education get smaller and those attending upper secondary education increase in size, the government could make more-efficient use of resources by not replacing retiring teachers in basic education and not increasing the teacher workforce proportionally to the number of students in secondary education, allowing class sizes to grow at least slightly.

Mexican students generally score low in international studies of academic achievement, not only when compared with developed nations but also within Latin America. Mexico placed last in the 2003 OECD Programme for International Student Assessment (PISA) mathematics scale (OECD, 2006b) and scored below the mean in a Latin American student assessment in 1997 (United Nations Educational, Scientific and Cultural Organization [UNESCO], 1998). These and other indicators—such as repetition and dropout rates—show that the overall quality of education in Mexico is very low (Andere, 2003; Guichard, 2005; Santibañez, Vernez, and Razquin, 2005).

National assessments of student learning, conducted in 2005 by the National Institute for Education Evaluation (Instituto Nacional para la Evaluación de la Educación, or INEE), also reflect the largely low achievement of Mexican students. Moreover, they highlight important disparities in achievement across different types of

[8] Some examples are Escuelas de Calidad, which gives competitive grants mainly for infrastructure improvement and encourages local decisionmaking; Enciclomedia, which digitizes primary-education textbooks so students can learn by using computers; and Red Escolar, a program that seeks to promote the participation of teachers and students in the use of new technologies by applying them to education (Santibañez, Vernez, and Razquin, 2005).

schools. Figure 10.4 shows that, in 2005, public urban schools had higher percentages of students in the medium and advanced levels of mathematics achievement than public rural schools and that public schools in general have significantly lower achievement than private schools.[9] Often overlooked, indigenous schools have by far the lowest achievement of all schools in Mexico, with less than 1 percent of their students getting placed in the advanced achievement level.[10] Between

Figure 10.4
Percentage of Mexican Sixth-Grade Students, by Mathematics Achievement Level and Type of School, 2005

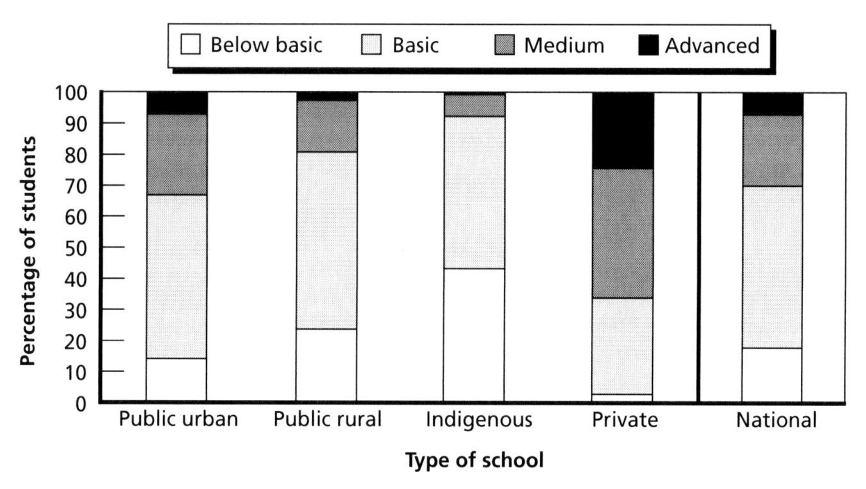

SOURCE: Robles Vásquez and Martínez Rizo, 2006.
RAND *MG985/1-10.4*

---

[9] Significant disparities can also be observed across Mexico's 32 states. Poor states, such as Chiapas and Michoacán, have almost 30 percent of students below basic achievement, while Distrito Federal has less than 9 percent of students in this level and more than 45 percent in either the medium or advanced level. For more information, see Robles Vásquez and Martínez Rizo (2006).

[10] Supplying education in indigenous communities represents a challenge to the public education system. According to the Comisión Nacional para el Desarrollo de los Pueblos Indígenas (CDI, or National Commission for the Development of Indigenous Peoples), in 2005, about 9.8 percent of the country's population was indigenous (0.7 percent lower than in 2000); moreover, the indigenous population often resides in the hardest-to-reach areas because only 17 percent of them live in one of Mexico's major cities (CDI, 2006). Primary education is provided to them in 43 different languages (Guichard, 2005).

2005 and 2007, the percentage in the categories signifying below basic achievement fell for all types of schools, with improvements in terms of percentage declines being particularly marked in public rural and indigenous schools (INEE, undated).

Mexico faces several challenges in its education system. Improving educational quality is obviously a major policy objective that requires improvements in infrastructure and teacher preparation. Another challenge is to increase enrollment and retention rates beyond basic education. Any policies that seek to address these issues must also take into account the existing achievement disparities across indigenous, public, and private schools, taking special care to prevent the most disadvantaged from being left behind.

Health

There has been a notable improvement in Mexican health standards during the past half-century; this is both a cause and a result of a changing demographic profile. However, inequality is still evident in health status and in access to health-care services, especially among those in poverty and those in rural communities. Mexican levels of spending in health are low when measured per capita or as a percentage of GDP, even compared with levels in other Latin American countries, and the health system appears to be significantly less efficient than those of other countries (OECD, 2005). Reforms during the 2000s sought to address these issues, but their impact and feasibility are yet to be evaluated.

Between 1950 and 2005, life expectancy in Mexico increased by almost 27 years; women can now expect to live almost 78 years and men almost 73 (Fundación Mexicana para la Salud [FUNSALUD], 2006). Infant mortality rates have steadily declined in the past five decades. Figure 10.5 shows that infant mortality in the general population declined from 40 deaths per 1,000 live births in 1990 to 16 in 2006.

Infectious diseases and malnutrition, which were among the main causes of death, have given way to nontransmissible diseases—such as cardiovascular diseases and diabetes—which, in 2009, accounted for more than 70 percent of all deaths (see Figure 10.6). This epidemio-

Figure 10.5
Life Expectancy and Infant Mortality, 1990–2010

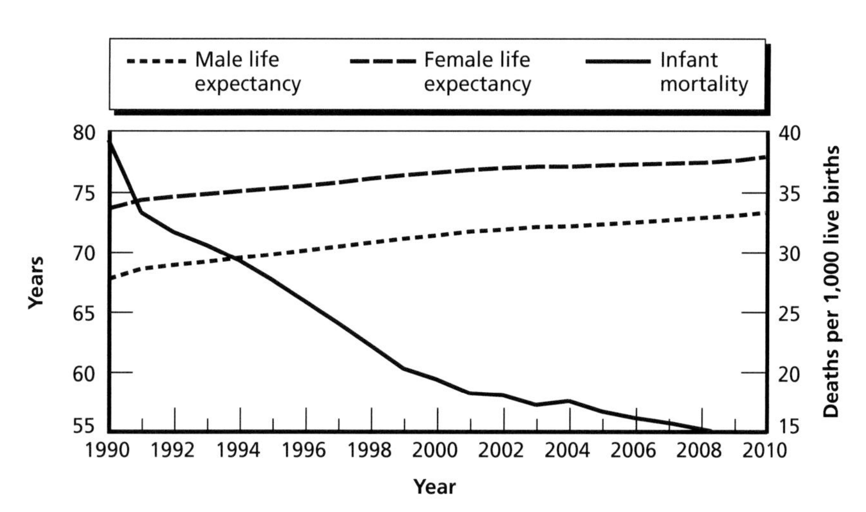

SOURCE: CONAPO, 2007.
NOTE: Figures beyond 2006 are projections.
RAND *MG985/1-10.5*

logical shift can largely be explained by—and is a cause of—Mexico's demographic transition: Although Mexico is still young compared with other OECD countries, the structure of the population has been aging rapidly as a result of decreases in fertility and mortality rates, and this trend is expected to continue for the next 50 years (OECD, 2005).

Despite the global decrease in mortality due to infectious and nutritional diseases, important inequalities exist in this epidemiological transition: Communicable diseases are still an important cause of death in rural and poor communities. For example, although noncommunicable diseases represent 77 percent of death causes in low-poverty areas, they account for only 57 percent of deaths in areas of high poverty. Similarly, the risk of death due to transmissible diseases or malnutrition is more than 30 percent higher in rural areas than in urban communities (Sistema Nacional de Información en Salud [SINAIS, or National Health Information System], 2008). A likely cause for these disparities is the unequal access to health-care services; rural areas have weaker access to health care, either because no services exist or because

Figure 10.6
Main Causes of Death in Mexicans, 1980–2009

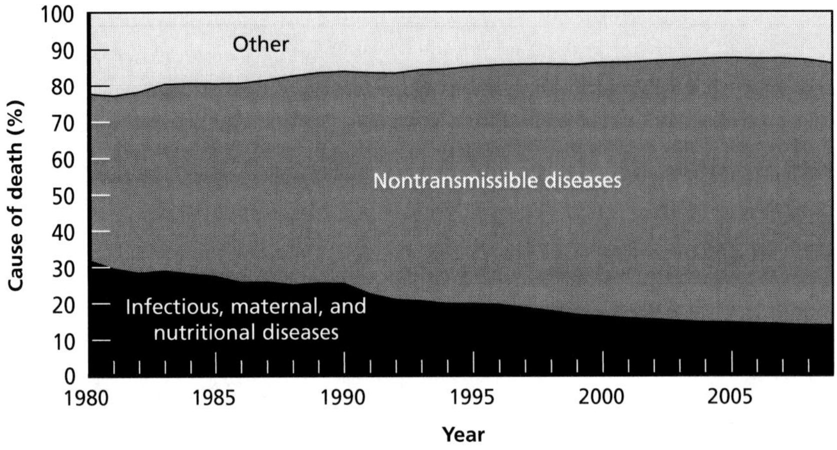

SOURCE: SINAIS, 2008.
NOTE: Figures beyond 2007 are projections.
RAND *MG985/1-10.6*

they are difficult to reach. Inequalities in access to health care also exist across states, with the richer northern states being much better served than states in the south (OECD, 2005).

Another cause for concern is the relatively low level of resources being allocated to the Mexican health system (FUNSALUD, 2006). Table 10.2 shows how Mexico's allocation of resources to health, measured by total health expenditure and government health expenditure, compares with resource allocations in other North American and Latin American countries. Clearly, there is a significant gap in health investment in Mexico when compared with these countries. Not only is expenditure on health low; some estimates suggest that resources are used less efficiently than in other countries. For example, administrative costs in Mexico represent a far bigger share of total health expenditure than in any other OECD country; this share is almost 40 percent larger than that of the United States, known for its high administrative costs (OECD, 2005).

The increasing importance of noncommunicable diseases, insufficient investment on health care, and the apparent need to redistrib-

Table 10.2
Total Expenditure and Government Expenditure on Health, Selected Countries, 2009

Country	Total Expenditure on Health as Percentage of GDP	Government Expenditure on Health as Percentage of Total Government Expenditure	Per Capita Total Health Expenditure ($) (PPP)	Per Capita Government Expenditure on Health ($) (PPP)
Mexico	6.5	11.9	846	408
United States	16.2	18.7	7,410	3,602
Canada	10.9	17.1	4,196	2,883
Argentina	9.5	14.6	1,387	921
Brazil	9.0	6.1	943	431

SOURCE: World Health Organization (WHO), undated.

ute health-care spending to improve efficiency motivated a reform to the general health law in 2003 (FUNSALUD, 2006). The main result of this reform was the creation of the System for Social Protection in Health, which is part of the ongoing effort to move the Mexican health-care system from a structure of vertically organized providers targeting specific populations toward a horizontally structured system closer to universal coverage (OECD, 2005).[11] Another effect of this reform is an increase in health spending, which was expected to reach 7 percent of GDP by 2010. This would represent an almost 8-percent increase over the 2005 level but still remains far from the spending levels of other countries in the region with similar national income (FUNSALUD, 2006).

These and other reforms to the general health law were implemented gradually between 2004 and 2007, so it is too early to evaluate their results. However, some researchers are already questioning the feasibility of the new regulations and pointing out their limited impact

[11] OECD (2005, Chapter Eleven) provides a comprehensive description of the Mexican health-care system and its structure. Chapter Three of that publication gives a detailed discussion of health system reforms and their objectives.

during the first year after the law was passed (Homedes and Ugalde, 2006). Moreover, even if the reforms were implemented successfully, the health system would still remain far from the government's long-term vision of a horizontal structure, delivering efficient health services with high quality (OECD, 2005).

Health Insurance and Social Security Coverage

Health insurance and social security coverage in Mexico are not universal and are highly fragmented. Workers in the private and public sectors are covered mainly by two social security agencies, IMSS and ISSSTE. These two institutions, combined with health services agencies for the armed forces and the national oil company (PEMEX), cover between 50 and 55 percent of the Mexican population. The rest of the population—the uninsured, self-employed, and workers in the informal sector—receive health-care services from the Ministry of Health and, in rural areas, a health services program for the population in extreme poverty called IMSS-Oportunidades (OECD, 2005). Another poverty-alleviation program that provides health care for the uninsured implemented by the federal government is called Seguro Popular. Seguro Popular has expanded substantially since its creation in 2001. The aim of this program is to provide access to health-care coverage for all Mexicans. Seguro Popular targets families in the lowest six deciles of the income distribution. In 2008, the program covered 20 million Mexicans (Aguila et al., 2011). Private insurers cover only about 3 percent of the population. Finally, the poorest segment of the population (about 3 percent) does not have geographic access to formal health-care facilities and relies mostly on traditional medicine for its health-care needs (OECD, 2005). For those individuals with no social security benefits, noncontributory pension programs have been available in Mexico since 2001. These programs provide a minimum flat pension for older persons, but they are still not universal (Aguila et al., 2011).

Among the main concerns for the government in terms of health-care provision and social security coverage are individuals in the informal sector—that is, those who pay neither taxes nor social security contributions. According to the OECD (2006c), the informal sector represents around 43 percent of total employment in Mexico. This constitutes a challenge in the labor market not only because of the reduced tax base and low productivity but also because health and social security provision are not guaranteed. The aim of Seguro Popular and noncontributory pensions is to provide coverage of health-care and social security benefits particularly for those in the informal sector. However, these programs are not yet universal; due to the aging population in Mexico, they could represent a high proportion of government budget in the near future. More policies to move workers from the informal to the formal sector should be promoted to lessen poverty in old age and increase available coverage of health-care and social security benefits through safety-net programs (Aguila et al., 2011). Some policies have been implemented to generate incentives for individuals to move from the informal to the formal sector. Among them is the 1997 Mexican pension reform, in which the traditional PAYG system was substituted by a fully funded system with PRAs. Mexican policymakers have argued that the PRA system is easier to monitor and that individuals perceive the individual account as their own saving (Aguila, 2011).

In economies with high labor turnover and migration between the formal and informal sectors, it is important to understand employment patterns. Marrufo (2001) finds movement from the informal to the formal sector as a result of the pension reform. Individuals with short spells in the formal sector might not achieve the minimum requirements to obtain a pension, so it is also important to understand working life in the formal sector. Aguila, Aguilera, and Velázquez (2008) find that the pension reform increased periods of contribution to the social security system in the formal sector. The latter might indicate that the pension reform provides incentives to move to the formal sector and remain in it. The reform had a greater impact in urban areas because most individuals in the formal sector reside in such areas. Therefore, this policy was most valuable in decreasing poverty levels in urban areas. Additional policies are needed to generate further movement to the formal economy in rural and urban areas.

Public Social Programs

Secretaría de Desarrollo Social (SEDESOL) is the Mexican ministry responsible for planning and coordinating the social policies of Mexico's federal government. SEDESOL's programs can be classified according to the goals outlined in the National Social Development Plan 2001–2006 (SEDESOL, 2001): (1) to reduce extreme poverty, (2) to create equal productive opportunities for the most-vulnerable populations, (3) to support the development of the capabilities of those in poverty, and (4) to strengthen social safety nets. Table 10.3 summarizes Mexico's social programs.

As of 2011, the largest poverty-alleviation program in Mexico is Oportunidades, which integrates education, health, and nutrition interventions while encouraging the active participation of all members of the family in improving their youth's education completion and the

Table 10.3
Objectives and Descriptions of Mexico's Social Programs, 2011

Objective	Program	Description
Reduce extreme poverty	Oportunidades	Largest poverty-alleviation program in Mexico Integrates education, health, and nutrition interventions while encouraging the active participation of all members of the family in improving youths' education completion and the health and nutritional status of the family Covers practically all Mexican households living in extreme poverty in both rural and urban communities A conditional cash-transfer program in which families receive monetary transfers conditional on their fulfilling their obligations, which include keeping their children in school and attending clinics for health education and clinical evaluations Well-known internationally due to its rigorous and independent evaluation system, which has highlighted the positive results in several areas and has encouraged other countries to implement similar programs
	Other programs	Subsidize milk for poor households Provide free and subsidized food and medicines in marginalized rural communities Give monetary support for higher-education students to engage in projects to increase development in marginalized communities

Table 10.3—Continued

Objective	Program	Description
Create equal productive opportunities	Opciones Productivas	Main program seeking to provide opportunities for Mexicans to engage in income-generating activities Allows individuals in the most-marginalized regions to develop their own productive projects and opportunities for self-employment, helping them increase their income and their families' well-being, as well as to have access to financial services, such as saving and lending
	Other programs	Provide beneficiaries in rural areas with transitory employment opportunities and training Support Mexican craftsmanship through (1) direct acquisition of products, (2) organization of regional crafting contests, (3) training of craftspeople, and (4) financing craft production Provide day care for mothers who work, study, or seek employment or for single parents with children 1–3 years old
Support capability development	Tu Casa	Reduces family vulnerabilities by increasing their wealth Combines families' own savings with subsidies for housing acquisition or improvement
	Other programs	Focus on rural and indigenous communities, providing families with subsidies to build, buy, or improve a home Advance the legalization of land ownership by expropriating land and either selling it to those who have illegally been living in it or making it available for urban or housing developments
Strengthen social safety nets	Program to Support Farming Laborers	Provides multiple benefits—housing, drinking water, social security, food, education, employment and training, safety, traveling assistance, and support during unexpected events and disasters—to farming laborers and their families Currently operates in 18 states
	Other programs	Support the poor elderly who are not beneficiaries of other federal programs, through cash transfers and nutritional education Serve all adults 60 or older, providing them with multiple services and support to contribute to their economic, health, and social development

SOURCE: Data on all programs, by ministry, from Presidencia de la República, undated.

health and nutritional status of the family. Oportunidades now covers practically all Mexican households living in extreme poverty, in both rural and urban communities. Oportunidades is a conditional cash-transfer program in which families receive monetary transfers conditional on their fulfilling their obligations, which include keeping their children in school and attending clinics for health education and clinical evaluations. The program is also well known internationally due to its rigorous and independent evaluation system, which has highlighted the positive results in several areas and has encouraged other countries to implement similar programs. In addition to Oportunidades, other poverty-reduction programs provide subsidized milk to poor households, free or subsidized food and medicines in marginalized rural communities, and monetary support for university students to engage in development projects in marginalized communities (review of data from Presidencia de la República, undated).

The main program seeking to provide opportunities for Mexicans to engage in income-generating activities is Opciones Productivas, which allows individuals in the most-marginalized regions to develop productive projects and opportunities for self-employment, helping them increase their income and their families' well-being. It also provides access to financial services, such as saving and lending. Another program provides beneficiaries in rural areas with transitory employment opportunities and training if it is required. Mexican craftsmanship is supported by the Fondo Nacional para el Fomento de las Artesanías, which manages four different programs to market crafts, organize contests, train craftspeople, and finance craft production (review of data from Presidencia de la República, undated).

Tu Casa's goal is to reduce family vulnerability by increasing their wealth, combining families' own savings with subsidies for housing acquisition or improvement. A different program focuses on rural and indigenous communities, providing families with subsidies to build, buy, or improve a home. The Comisión para la Regularización de la Tenencia de la Tierra operates two other programs to advance the legalization of land ownership by expropriating land and either selling it to those who have illegally been living in it or making it available for urban or housing developments (review of data from Presidencia de la República, undated).

The first of the programs that focus on vulnerable populations provides multiple benefits—housing, drinking water, social security, food, education, employment and training, safety, traveling assistance, and support during unexpected events and disasters—to farming laborers in 18 states. Two other programs focus on the older population; the first of them supports the poor elderly (those who are not beneficiaries of other federal programs) through cash transfers and nutritional education. The other serves all adults 60 or older, providing them with multiple services and support to contribute to their economic, health, and social development (review of data from Presidencia de la República, undated).

Mexico has a long tradition of instituting social policy programs. Poverty remains an important problem in Mexico; therefore, the design of social programs, as well as evaluation of their effectiveness, is important. The evaluation of programs, such as Oportunidades, has allowed better targeting of social policies. However, the effectiveness of many other programs remains largely unknown.

Remittances

Three types of financial flows are generated from international migration: pension benefits to previous migrants now residing in Mexico, wages to Mexican border residents who work in the United States, and money flows transferred to Mexico by Mexicans residing abroad, also known as remittances (Shaffer, 2004). Of these, remittances are by far the most important to Mexico, the second-largest remittance recipient in the world after India (Orozco, 2003). Figure 10.7 shows BANXICO's estimates of total remittance flows and number of remittance transactions; both increased considerably in the late 1990s through the mid-2000s. Remittances grew from $4 billion in 1996 to more than $26 billion in 2007 before dropping to $21 billion in 2010 (BANXICO, undated).

In the past, remittances and FDI were considerably lower than FDI (traditionally, the most important source of external funding for developing countries). However, remittances and FDI were virtually

Figure 10.7
Remittance Flows to Mexico and Number of Transactions, 1996–2010

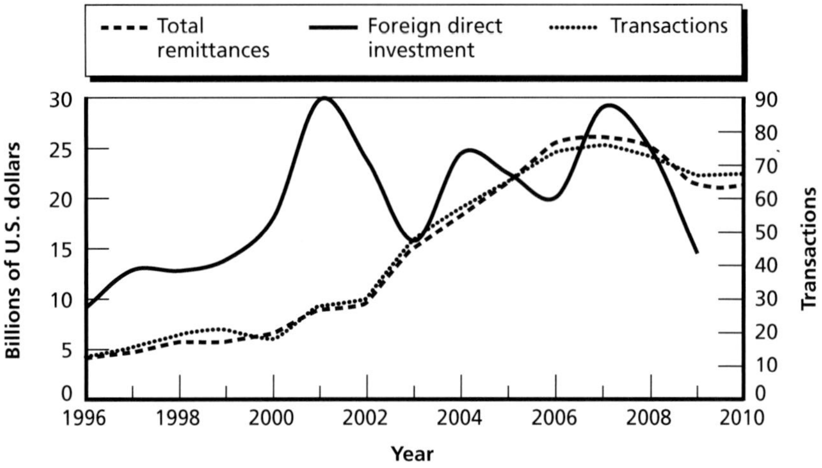

Total remittances — Foreign direct investment — Transactions

Billions of U.S. dollars: 0, 5, 10, 15, 20, 25, 30
Transactions: 0, 10, 20, 30, 40, 50, 60, 70, 80, 90
Year: 1996, 1998, 2000, 2002, 2004, 2006, 2008, 2010

SOURCE: BANXICO, undated.
RAND *MG985/1-10.7*

identical in 2005 and 2007, and remittances actually surpassed FDI in 2006 and 2008. Nevertheless, the explosive growth in remittances has slowed down in parallel with the global economic crisis, exhibiting virtually no change between 2006 and 2007, declining slightly between 2007 and 2008 and more drastically between 2008 and 2009, after which it stabilized until 2010 (BANXICO, undated).

Although these figures underscore the swift increase of remittance flows between 2000 and 2006, it should be noted that remittance estimates from other sources differ considerably from official reports. For example, remittance estimates obtained from survey instruments are as much as 50 percent lower than those from BANXICO. Some attribute at least part of the difference to the fact that survey data are collected during the periods of the year when remittance flows are smaller, while others argue that official figures overestimate remittance flows because they do not correspond to remittances and include other sources of transfers, perhaps even illegal activities, such as money laundering (Zárate-Hoyos, 2005; García Zamora, 2005a; Tuirán Gutiérrez, Santibáñez Romellón, and Corona Vázquez, 2006).

On the other hand, it has also been argued that official estimates might not include a significant portion of total remittance flows because as many as 30 percent of Mexicans who receive them do so through informal channels, such as messengers, ordinary mail, or directly when migrants visit their home communities (Inter-American Development Bank [IADB], 2003).[12] In short, any estimates of remittance flows must be interpreted with caution. Regardless of these shortcomings, the BANXICO estimates are the only time-series data available for remittances in Mexico and the only ones that allow for international comparisons.

In addition to estimates of total remittances, household surveys provide information on the number of households receiving remittances and their characteristics. These studies indicate that only between 4 and 6 percent of households in Mexico receive remittances. Although this represents only a small fraction of all Mexican households with migrants, remittances appear to be extremely important to these households for several reasons, including the following: (1) Remittances represent, on average, almost 40 percent of their income and more than 50 percent in the top three recipient states; (2) four out of ten remittance-receiving households do not receive income from other household members; (3) seven out of ten remittance recipients are female; and (4) receiving households have a higher ratio of economically inactive to economically active people than households that do not receive remittances (Canales, 2004; Zárate-Hoyos, 2005; Shaffer, 2004).

In recent years—mainly because of increased awareness of the economic magnitude of remittance flows to developing countries—governments and international organizations have become more interested in remittances' potential to be a tool for economic development (Moctezuma, 2006), and it has been suggested that migration and remittances reduce poverty in developing countries (R. Adams and Page, 2005). Institutions, such as the World Bank and the IADB, have organized studies and conferences exploring policies to channel

[12] Zárate-Hoyos (2005) points out that nonmonetary transfers in the form of goods and services sent or brought to Mexico by migrants represent a possibly significant source of remittances that is not included in BANXICO's definition.

remittances to productive projects and promote the economic development of the communities that receive them.[13] Federal, state, and local governments have also paid more attention to remittances after some states' positive experiences creating programs to channel remittances from migrant organizations toward financing infrastructure, public services, recreation facilities, and other community-related projects.

An important example is the Tres por Uno (Three-for-One) program, which began in the state of Zacatecas in 1993. In this program, migrant organizations send remittances, and each dollar sent is matched by the federal, state, and municipal governments. In 2002, Tres por Uno officially became a federal program available to all states in the country but with a limited yearly budget for which states compete. In 2005, for example, Tres por Uno invested a total of MXN$857 million—around US$78 million—in projects. Although 26 states participated in the program in 2005, there is clearly an advantage to being a traditional migrant-sending state because Jalisco and Zacatecas accounted for more than half of the financed projects between 2002 and 2005 (Soto Priante and Velázquez Holguín, 2006).

Another recent line of research is the examination of how families use the remittances they receive and how the remittances affect their well-being. For example, Amuedo-Dorantes, Sainz, and Pozo (2007, p. 23) find that remittances have a greater impact on health-care expenditures than other sources of income do, indicating that remittances might "play a crucial role in supplementing any deficiencies in the public provision of medical services."

[13] A good starting point to learn more about these publications and projects is IADB (undated).

CHAPTER ELEVEN

Conclusions and Policy Recommendations

Fundamental reform proceeds slowly in a democracy with a divided government. Mexico's multiparty presidential democracy will improve policy transparency and reduce governmental corruption in the long run. However, in the short run, the divided government has made it difficult to gain approval for substantial economic and social reforms. Given these constraints, the strides Mexico has made on economic and social policy reforms are all the more impressive.

To improve the lives of Mexican citizens on a day-to-day basis, as well as give surety to Mexico's economic and social success, Mexican legislators will need to set priorities and strategies for further, more-aggressive reform. Because Mexico represents the largest source of immigration to the United States and this flow has important implications for the United States, U.S. policymakers can assist in prioritizing and supporting policy changes. Four potential priorities, based on specific areas of concern, emerged from our study. We describe these in this chapter.

Economic Policy Options

Our study found that, although Mexico was subjected to many adverse economic situations in the 1980s, relief came in the mid-1990s through a combination of moves toward the privatizing of government companies, reducing the VAT for many goods, and significantly decreasing income taxes. Additionally, NAFTA allowed Mexico to continue its change from a primarily closed economy to an open one.

We recommend that the following policy options be considered for further improvement in this area:

- *The CFC should be vested with real powers to prevent predatory pricing, divest monopolies, and implement leniency programs* (World Bank, 2006a). So far, the CFC has had very limited power to prevent and take action against monopolistic practices. The CFC urgently needs to become a solid organization that strongly promotes market competition.
- *The tax base could be broadened in order to improve government revenues and capability to target social and economic issues.* Tax collection can be improved by providing incentives to states for collecting taxes and for firms to move to the formal economy.

Energy Policy Options

Our findings suggest that it might be the appropriate time to dramatically transform the Mexican energy sector because indicators suggest that government revenues have an unhealthy dependence on oil. We recommend that the following efforts be considered:

- *Allow private producers into some segments of Mexico's energy sector.* In 1992, when the need to provide adequate electrical power to sustain growing electricity demand became clear, the Mexican Congress amended the constitution to attract private investment and allow limited private generation of electric power. The telecommunication sector was successfully privatized in 1990. Similar moves might be made in the energy sector, especially in exploring and extracting oil from the hard-to-extract fields.
- *Invest in expanding refining capacity and natural-gas production.* Most of the production and exploration of energy from PEMEX, the government monopoly, is under severe pressure to update and improve efficiency. Allowing foreign participation can provide a rapid solution to these issues and improve government revenue.

Labor Policy Options

Our research into the state of Mexican labor policy suggests that it is time to build on reforms enacted recently in order to meet new 21st-century challenges. Early in his presidency, President Obama called for stronger labor rules enforceable within NAFTA ("Obama Hopeful of Fixing Truck Dispute with Mexico," 2009). Developing greater flexibility in Mexican labor markets need not be incompatible with the U.S. president's call if the overall aim is to provide better working conditions for Mexican labor. Several steps might be taken in this direction:

- *Engender growth in the Mexican formal economy.* Working conditions would improve for many more citizens if workers were able to move from the informal to formal economy, because there are better retirement plans and medical coverage for workers in the latter. Policies to retain individuals in the formal sector are hard to implement, but programs to generate employment and economic growth and to improve country comparative advantages are key to improving conditions in the labor market.
- *Effectively restructure labor regulation to meet workforce needs.* Labor regulation in Mexico is out of date, and overly stringent regulations contribute to the size of the informal sector. Lack of flexibility for part-time jobs and reentering the formal labor market after retirement are some of the issues that affect the most vulnerable segment of the population.
- *Allow for less costly hiring and firing practices by firms in Mexico.* By doing so, the process of innovation that depends on quick formation and dissolution of firms will be aided while still adhering to rules of due process.

Other Social Policy Options

Mexico has a wide range of social programs to alleviate poverty. However, other than Oportunidades, the programs' individual impact in reducing poverty has not been fully evaluated. We recommend that the following policy options be considered to overcome widespread social and economic inequality in Mexico:

- *Coordinate social policy across multiple sectors—health, education, social security, and poverty alleviation.* It seems that many social programs have had a low impact in reducing poverty, and their specific objectives often overlap. An effective and efficient social policy would require eliminating programs that have overlapping objectives and concentrate efforts on those with the greatest impact.
- *Make educational quality a primary policy objective as soon as possible.* Mexico faces several challenges in its education system. Improving education quality is obviously a major policy objective that requires improvements in infrastructure and teacher preparation. Another challenge is to increase enrollment and retention rates beyond basic education. Any policies that seek to address these issues must also take into account the existing achievement disparities across indigenous, public, and private schools, taking special care to prevent the most disadvantaged from being left behind.
- *Promote economic growth in rural areas.* The OECD (2007b) finds that it is crucial to have more coordination among the institutions in Mexico responsible for rural policies. A better arrangement would allow institutions in charge of rural development to exploit synergies and identify potential complementarities in their policies.
- *Reform those pension systems with cash-flow deficits.* Pension systems represent an important government liability. The private-sector social security system (IMSS) was reformed in 1997, and ISSSTE (the counterpart for government employees) was reformed in 2007. Other public-sector institutions, such as the oil and electricity companies, universities, and many local governments, need to pay attention to reform their pension systems and guarantee their future financial feasibility. Reducing the government burden

from the pension systems could allocate more resources to policies that promote economic growth.

PART THREE

The Past and Present of U.S.-Mexican Relations

To shed light on the state of U.S.-Mexican relations and to examine the issues on which greater understanding between Mexico and the United States might be facilitated and improved, we conducted further analysis in additional areas.

In this third part, we address the following questions:

- What has been the fate of immigration reform in both countries?
- What is the prevailing opinion on Mexican immigration among the American public?
- How do Americans view NAFTA?

Schaefer, Bahney, and Riley (2009), in a companion report to this volume, examine the security situation in Mexico and assess its impact on the United States. The authors outline policy options that the United States can consider in assisting Mexico in its efforts to improve internal security. Our goal is to build on this important work by bringing to light further topics critical to current and future U.S.-Mexican relations: the state of NAFTA and U.S. public opinion about the agreement.

This third part is divided into three chapters. The first chapter discusses immigration reforms proposed in the past decade. The second chapter takes a look at current immigration and NAFTA policy in light of U.S. public opinion; finally, the third chapter concludes with a discussion of the primary trends in these areas and several policy options that might improve U.S.-Mexican relations in the future.

CHAPTER TWELVE

Immigration Policies and Proposals During the 2000s

Immigration Policy Proposals in Mexico

After the terrorist attacks of September 2001 changed the bilateral agenda on migration, negotiations between Mexico and the United States on migration policy effectively stalled. In February 2006, the Mexican Congress—both the Senate and Chamber of Deputies—unanimously passed a resolution titled "Mexico and the Migration Phenomenon" (Embassy of Mexico, 2006), which presented general principles and recommendations to guide future migration policies; one of the most-important points made in this document is the public acknowledgment that *the migration phenomenon is a shared responsibility of both Mexico and the United States, so both countries, not only the United States, must be part of the solution.* The document states that Mexico must improve and enforce its own migration laws, with full respect for the human rights of migrants and their families, and strengthen the fight against criminal organizations specialized in migrant smuggling and document falsification. As for a migratory agreement between the United States and Mexico, the document supports a far-reaching guest-worker scheme and the establishment of a bilateral medical insurance system to cover migrants and their relatives, as well as a totalization agreement of pension benefits. The latter would eliminate dual social security taxation for workers who are required to pay social security contributions in their country of origin and the country in which they are currently working, both from the same earnings. Additionally, a totalization agreement would allow workers who have employment histories in two countries but have not worked long enough in either country to qualify for pension benefits to combine work credits from both countries in order to be eligible for social security benefits (Social Security Administration [SSA], 2004).

Mexico's 2007–2012 National Development Plan includes among its main objectives "to protect and actively promote the rights of Mexicans abroad" and "to build a new migration culture." Among the strategies to achieve the first goal are (1) strengthening the network of Mexican consulates and (2) fortifying the linkages with the Mexican community abroad, particularly in the United States. The strategies to achieve the second objective involve mainly the respect and protection of immigrants from other countries in Mexico and the creation of economic opportunities to discourage migration, particularly in the regions that have traditionally sent migrants abroad.

Immigration Policy Proposals in the United States

The United States has a diverse collection of politically active groups seeking to influence many aspects of immigration policy and a long history of admitting large numbers of foreigners as permanent residents. Nongovernmental organizations, both those favorable to increased immigration and others that support a more restrictive policy, vie for influence in shaping U.S. immigration policies. The political process translates these conflicting goals into policies, and the 2000s were active years for immigration legislation. Three major U.S. migration policy options have been debated: guest workers, earned legalization, and legalization (Hanson, 2006).

Immigration Proposals Since 2000

Senator Larry Craig introduced the Agricultural Job Opportunity, Benefits, and Security Act of 2003 (S.1645), an earned-legalization program for farm workers that was endorsed by a majority of senators. However, the proposal was not enacted.

In January 2004, President George W. Bush unveiled a Fair and Secure Immigration Reform proposal (White House, 2004), which would permit the 6 million to 8 million unauthorized foreigners in the

United States with jobs to become legal guest workers. However, the Bush plan was not transformed into legislation. In the past few years, the United States has seen other failed immigration bills, as shown in Table 12.1.

The last bill on comprehensive immigration reform discussed in the U.S. Congress was the Comprehensive Immigration Reform Act of 2007. The bill, introduced in the Senate on May 9, 2007, failed by a vote of 46–53 on June 28, 2007; had it passed, the bill would have provided legal status and a path to legal citizenship for the approximately 12 million illegal immigrants residing in the United States at the time. The bill was portrayed as a compromise between legalization of illegal immigrants' status and increased border enforcement. For instance, it included funding for 200 miles of vehicle barriers, 70 camera and radar towers, and 18,000 more Border Patrol agents while restructuring visas around high-skill workers.

Table 12.1
Recent Immigration Proposals

Date	Proposal
September 2003	Agricultural Job Opportunity, Benefits, and Security Act of 2003, a bill proposed by Senator Craig (S.1645)
January 2004	Fair and Secure Immigration Reform proposal from President Bush (White House, 2004)
May 2005	Secure America and Orderly Immigration Act, a bill proposed by Senator John McCain (S.1033)
July 2005	Comprehensive Enforcement and Immigration Reform Act of 2005, a bill proposed by Senator John Cornyn (S.1438)
May 2006	Comprehensive Immigration Reform Act of 2006, sponsored by Senator Arlen Specter and passed in the Senate but not in the House (S.2611)
May 2007	Comprehensive Immigration Reform Act of 2007 sponsored by Senator Harry Reid (S.1348)
November 2010	Development, Relief, and Education for Alien Minors (DREAM) Act proposed by Senator Richard Durbin (S.3992)

Other key elements of the proposed bill were as follows:

- A new class of visa, the Z visa, would be given to everyone who was living illegally in the United States on January 1, 2007; this visa would have given its holder the legal right to remain in the United States for the rest of the holder's life and access to a Social Security number. After eight years, the holder of a Z visa would have been eligible for a green card.
- The employer-sponsored component of the immigration system would have been eliminated and replaced with a point-based merit system, with points based on a combination of education, job skills, family connections, and English proficiency. Additional points were to be awarded if a U.S. job offer was available. Several family-based immigration categories would have also been folded into the point system.

The bill received heated criticism from both sides of the immigration debate. Those who support tougher positions against immigration rejected providing amnesty for illegal aliens because it would reward them for disregarding U.S. immigration laws. On the other hand, immigration proponents criticized as unfair the point system and provisions limiting family-reunification visas available to only nuclear family members of U.S. citizens. Labor unions, human rights, and some Hispanic organizations attacked the guest-worker program, claiming that it would create a group of underclass workers with no benefits (Dinan, 2007; McGee, 2007; Dinan and Lengell, 2007).

With the failure to move comprehensive reform forward, there has been one other attempt at smaller reform, the DREAM Act. The DREAM Act would have given minor students who are illegally in the United States and graduate high school a chance to legalize their status (Bruno, 2011). Introduced in one version as early as 2001, the act was a response to a provision in the 1996 Illegal Immigration Reform and Immigrant Responsibility Act that discouraged states and localities from granting certain benefits for postsecondary education—such as in-state residency status for purposes of reduced tuition—to illegal immigrants. The U.S. House of Representatives approved DREAM

Act language in an unrelated bill in 2010, but the measure failed in the Senate. Another version was introduced in both the House (H.R. 1842) and Senate (S.952) on May 11, 2011; however, as of this writing, neither had proceeded to a final vote.

Given these developments, as of December 2011, the active policy on illegal immigration continues to be based on the IRCA, discussed in Part One. These policies are outdated because they reflect the debate and concerns of the 1970s and 1980s. In order to move forward, it is critical that policy be revised to reflect the current economic and political situation.

CHAPTER THIRTEEN

U.S. Public Opinion on Immigration and the North American Free Trade Agreement

Public opinion is a vital, dynamic force in supporting progress for international policies, institutions, and organizations, making it important to assess foreign policy issues from a public opinion perspective. In this chapter, we examine the attitudes that the American public has toward the political and economic effects of U.S.-Mexican relations following the passage of NAFTA.

Fluctuating U.S. Opinion on Immigration

Recent public opinion polls suggest that most Americans hold generally positive views regarding immigration. According to a Gallup poll conducted in June 2008, 64 percent of Americans think that immigration is good for the United States. This number is a slight increase from the 60 percent recorded in the June 2007 Gallup poll (J. Jones, 2008). Some experts suggest that the reason for reduction in anti-immigration sentiment is the shift in Americans' focus toward the struggling economy and rising energy prices. However, in 2009 and 2010, the percentage who thought that immigration was good for the United States dropped again to 58 and 57, respectively (Morales, 2010). At the beginning of 2008, 11 percent of Americans said that immigration was the most important issue that the country was facing. By June 2008, only 4 percent of Americans believed it to be a critical issue; even though the view that it was the most important topic in the country became more prevalent again, by July 2010, still only 7 percent mentioned it as the most critical topic (Newport, 2010). Only those Americans living in areas with more immigrants rank immigration as a bigger community problem, according to the survey conducted in early 2006 by the Pew Hispanic Center and the Pew Research Center for the People and the Press. Immigration emerges as a dominant local concern only in Phoenix, near a major entry point for illegal immigrants, where 55 percent say it is a very big problem.

At the same time, Americans still seem somewhat supportive of limits on the amount of immigration. According to the same poll, 39 percent of respondents said that the number of immigrants coming to the United States should be decreased. This is substantially lower than the 58 percent reached shortly after the September 11, 2001, terrorist attacks. Of the major U.S. racial and ethnic groups, Hispanics tend to be the most supportive of immigration, according to the Gallup June 2008 poll. Only 28 percent of Hispanics favor decreased immigration, compared with 39 percent of blacks and 42 percent of whites. The same poll finds that Republicans are more likely to favor decreased immigration than are Democrats.

The 2008 Gallup poll also finds that Americans hold negative views on immigration's effects on the economy: Sixty-three percent of the public said that immigrants cost taxpayers too much by using government services. On the other hand, 79 percent of Americans see some economic benefit in that, they believe, illegal immigrants tend to take low-paying jobs that Americans do not want. When asked a similar question in the Pew Hispanic Center/Pew Research Center for the People and the Press poll in early 2006, 65 percent said that immigrants coming to the country take mostly jobs that Americans do not want, rather than take jobs away from Americans.

The 2008 Gallup poll also finds that Americans are more favorable toward immigration from certain parts of the world than others. A large proportion of Americans (48 percent) think that there are too many immigrants from Latin American countries, 20 percent think that there are too many immigrants from European countries, and only 19 percent think that there are too many immigrants from African countries. The current state of immigration has its roots in the Immigration and Nationality Act of 1965 (Pub. L. 89-236), which not only increased the rate of legal immigration but also produced a dramatic shift in the immigrants' country of origin. Indeed, if, prior to 1965, two-thirds of legal immigrants to the United States were from Europe and Canada, more than half now come from Asia and Mexico (Kellam and Vargas, 1998; U.S. Census Bureau, 2011b).

Most polling organizations conclude that American views of immigration seem to vary largely depending on the strength of the domestic economy and, to a lesser extent, on prevailing racial and ethnic prejudices and perceived external threats. Except for the poll taken almost immediately after 9/11, Gallup data indicate that public views on immigration at least since the 1990s are significantly influenced by perceptions of the national economy. For example, extended support for limiting immigration coincided with heightened public concern about the recession of the early 1990s.[1] Similarly, the strongest showing in favor of maintaining or increasing immigration levels paralleled very favorable public impressions of the economy.

Illegal and Legal Immigrants

An important factor to consider is that polls usually ask about "immigration" in general. If respondents are asked specifically about "illegal immigration," responses are significantly more negative. For example, a CNN/Opinion Research Corporation poll conducted in January 2008 found that 65 percent of respondents would like to see the number of illegal immigrants currently in the country decreased.

Most Americans say that immigrants contribute to the country. In a May 2007 CBS News/New York Times poll, when asked without reference to immigrants' legality, a majority of Americans said that immigrants contribute to this country and work as hard as or harder than people born here. However, illegal immigrants are viewed particularly negatively in terms of their economic effects: Seventy percent think they weaken the economy because they use public services but do not all pay taxes.

---

[1] In general, U.S. public opinion on immigration (and trade) tends to be volatile, depending on the state of the economy and the tenor of the debate in Congress, among other factors.

The American public is deeply divided over how to handle illegal immigrants already in the United States, who are estimated to number about 12 million. According to the 2006 Pew Hispanic Center/Pew Research Center for the People and the Press survey, 53 percent say that people who are in the United States illegally should be required to go home, while 40 percent say that they should be granted some kind of legal status that allows them to stay in the United States. The same poll finds a division of opinion over how to stem the flow of illegal immigrants across the Mexican border. When asked to choose among three options, roughly half of Americans (49 percent) say that increasing the penalties for employers who hire illegal immigrants would be most effective in reducing illegal cross-border immigration, while one-third prefer boosting the number of Border Patrol agents. Only 9 percent of the public say that the construction of more fences along the Mexican border would be most effective.

U.S. Public Opinion on the North American Free Trade Agreement

In several polls conducted in the 2000s, Americans have generally expressed positive views on NAFTA. In a June 2005 international policy attitudes poll, 46 percent of respondents viewed NAFTA as being good for the United States, while 40 percent saw it as bad ("International Trade," undated). By 2008, the view had become somewhat more negative, with only 37 percent seeing NAFTA's impact on the U.S. economy as mainly positive and 53 percent stating that it was mainly negative (English, 2008).

Despite its general support for NAFTA, the U.S. public views Mexico to be the winner under this agreement. In a 2005 Ipsos Reid poll, 47 percent of Americans saw the United States as a loser and 43 percent as a winner. When the Chicago Council on Global Affairs surveyed the U.S. public in July 2004 about the impact of NAFTA, 78 percent of Americans saw the agreement as good for the Mexican economy but seemed divided about the benefits for the U.S. economy:

Only 42 percent saw NAFTA as good, and 43 percent saw it as bad for the U.S. economy.

Americans seem to subscribe to the overall benefits of free trade as represented by NAFTA but, at the same time, feel concerned about members of American society put at risk by freer trade. Fifty-four percent believe that FTAs take jobs away from Americans, while only 23 percent believe that U.S. jobs are created, according to a Rasmussen Reports survey conducted in June 2008. More than half of respondents think that NAFTA needs to be renegotiated, according to the same survey. The survey finds that 56 percent of respondents support renegotiation, and 39 percent say that U.S. FTAs in general have directly affected their families. Attitudes appear to have shifted only little in the years since 2008. A Rasmussen survey in 2010 found similar percentages, although, for example, the number of people who believed that FTAs took jobs away decreased to 45 percent, and 28 percent believe that U.S. jobs are created (Alvarado Nieto and Carbajo, 2010). In a separate Rasmussen Reports survey taken in 2008, it was found that Republicans narrowly believe that free trade is good (41 percent to 33 percent), Democrats narrowly believe that it is bad, and unaffiliated voters are evenly divided (Rasmussen Reports, 2008).

But when asked what, in principle, should happen to the level of trade between the United States, Mexico, and Canada, only a small minority favors reducing it. Given three options in an Ipsos Reid poll conducted in April 2005, only 8 percent felt that the trade and integration of the economies of these three countries should be reduced. Rather, 47 percent said that trade among these countries should stay the way it was at the time for the foreseeable future, and 41 percent said that the economies should be further integrated. Thus, a fairly strong majority believes that the United States should, at a minimum, continue trading with these countries.

Upon completion of its latter June 2008 survey, Rasmussen Reports found that, with rising oil and food prices and other growing economic stresses, it was perhaps not surprising to see respondents taking a harsher view of free trade than they had a year earlier.[2]

---

[2] The caveat mentioned in the section on U.S. public opinion on immigration applies here, too. Public opinion on trade tends to be volatile, depending on the state of the economy and other factors.

CHAPTER FOURTEEN

Conclusions and Policy Recommendations

Recent policy developments and public opinion present the underlying complexity of the present and future conditions of U.S.-Mexican relations. Immigration reform is actively debated in the U.S. Congress, but all initiatives since the IRCA passed have failed to pass. NAFTA has increased trade and investment between the two countries, but the issue of whether Mexican trucks should be given free access to the United States was solved only in 2011, more than ten years after such access was meant to be implemented.

U.S. public opinion is divided on whether illegal immigrants should be deported or granted legal status. Although most Americans want the number of illegal Mexican immigrants entering the United States to decrease, many feel that immigration is generally good for the U.S. economy. Correlatively, U.S. public opinion remains divided on whether NAFTA is good or bad for the U.S. economy. Although U.S. citizens subscribe to the philosophy of free trade overall, more than half would like to see NAFTA renegotiated.

Mexico, too, expresses political indecision that ultimately impedes international cooperation. Some critics viewed President Vicente Fox as partnering too closely with the United States and thus turning his back on Latin America. Indeed, in the period before he took office, President-elect Calderón said he would put a more Latin American face on Mexico. After a visit with President Michelle Bachelet of Chile, Calderón said that his government would look clearly toward the south because he would be a president with a deep Latin American conviction (Lund, 2006).[1]

Despite the ambivalence of both countries, however, Mexico is bound to the United States through trade, investment, people, and even culture. Even with continuing unease about the economic power of the United States—with a GDP more than 14 times that of Mexico, GDP per capita more than five times as large, and a population almost three times as large—Mexico's politicians see closer economic relations as good for their country. In a survey before the 2006 election, nearly all Partido Acción Nacional (PAN) candidates for the Chamber of Deputies said they favored expanding commercial ties with the United States. Interestingly, the vast majority of Partido de la Revolución Democrática (PRD) candidates also wanted to expand commercial ties, with only 22 percent wanting to maintain them at current levels or reduce them. This suggests a broad agreement that Mexico's economy depends heavily on the economy of the United States (Bruhn and Greene, 2007).

It seems fair to say that—at least on the immigration and economic fronts—deterioration and improvement both seem distinct possibilities in the relations between the two countries. Possibilities and suspicions abound. Policies adopted by current and future U.S. and Mexican administrations are crucial in determining which future direction these relations finally take.

Improving Outcomes with the North American Free Trade Agreement

A policy option that could be implemented with NAFTA in place is to *explore opportunities to geographically expand U.S. investment in Mexico.* Much of the FDI flowing into Mexico has been concentrated in the

---

[1] As quoted in Lund (2006), Calderón said, "Mi gobierno va a mirar claramente hacia el sur porque soy un presidente con una profunda convicción latinoamericana" [my government will clearly look to the south because I am a president with a deep conviction in Latin America].

six states along the U.S. border and Mexico City.[2] In order to foster more-even development, Mexican policymakers should encourage the setting up of enterprises in other locations to take advantage of local comparative advantages. A sound transportation infrastructure can mitigate the increased cost of being located away from the border.

A Public Opinion Policy Option

There is also a policy option in addressing public opinion: *Improve international understanding through media and educational forums.* Improvement of U.S.-Mexican relations begins with a strong foundation of public understanding on both sides of the border. Presenting information on Mexican and American culture, society, and politics via meaningful media outlets and formal and informal instructional opportunities in both countries will strengthen this critical relationship by educating citizens about past, present, and future international policies and concerns, especially if the forums of one country demonstrate an understanding of the concerns of the other.

---

[2] For example, FDI flows into production plants called maquiladoras that were originally linked to U.S.-based plants and generally were foreign invested. After creating the maquiladora program, Mexico, in 1990, created a similar program for domestic producers, called the Program for Temporary Imports to Promote Exports (PITEX) (Cañas and Gilmer, 2007). In 2007, the two programs were merged under the Decree for the Promotion of the Manufacturing Industry, Maquiladora, and Exportation Services (Decreto para el Fomento de la Industria Manufacturera, Maquiladora y de Servicios de Exportacion, or IMMEX program) (van't Hek, Becka, and Mejia, undated). According to Brauer (undated), a maquiladora company is a Mexican corporation operating under a special customs treatment, whereby it may temporarily import into Mexico machinery, equipment, replacement parts, raw materials, and, in general, everything needed to carry out its production activities.

PART FOUR

Conclusion

CHAPTER FIFTEEN

Conclusion

The findings and observations discussed in this monograph demonstrate that Mexican migration to the United States and the economic and social conditions and policies in Mexico must be reconsidered in a binational framework. Expanding cooperation on important issues concerning trade, immigration, security, and business investment has the potential to advance the prosperity and security of both countries. U.S. opinion polls suggest that U.S. citizens have varying and frequently negative attitudes toward Mexico and toward U.S.-Mexican relations in particular. The findings in this monograph suggest ways in which the two countries can further come together, from instituting organizations that facilitate immigration in ways that are valuable for both countries to building policies and programs to assist Mexican citizens who stay in Mexico.

This final chapter summarizes our quantitative and qualitative data and offers concrete recommendations for policies and programs affecting Mexican and U.S. immigrants and Mexican citizens.

The first part of this monograph described the conditions precipitating the large immigration flow of Mexican citizens to the United States, especially during the 1990s and early part of the 21st century. Existing quantitative data indicate that, compared with the 20th century, the population of Mexico–to–United States immigrants is increasing but also changing. The number of Mexican immigrants in the United States has risen from 2.5 million to 11 million in the past 26 years, and the group has traditionally been made up of educated men seeking to diversify sources of household income. However, as the poorest regions in southern Mexico begin to feed more into migratory flows, it is less likely that these immigrants will have received the same amount of education as their predecessors, thus offering a different skill set to their receiving nation.

Mexico is not considered a poor country; by global standards, it is a middle-income nation (World Bank, undated [a]). Some reports have even suggested that there will be labor shortages developing in many parts of the country because of a rapidly aging population, declining fertility rate, and a constant migration flow. Despite such projections, however, Mexican citizens seeking to move to the United States are still typically driven by personal economic factors. Our examination suggests that there is still a large wage differential between the two countries and that many areas in Mexico still suffer from economic crises promulgated by lack of industry and crop failure in rural areas. Numbers of Mexican immigrants to the United States are expected to increase also because of expanding social networks. Current Mexican workers in the United States are often asked to locate additional workers for their employers. This phenomenon can ease the migratory process for new legal immigrants because their ties assure them of a higher likelihood of employment than other, nonconnected Mexican citizens.

Mexican immigration has a considerable effect on the U.S. population. It is important to keep in mind that, although the numbers of Mexican immigrants are gradually increasing, many migrants never actually return to their original country or they stay for longer periods than originally planned. We suspect that this trend will increase the number of Mexican-born citizens staying in the United States, and this, of course, has implications for the U.S. economy. Some studies suggest that U.S. citizens ultimately benefit from Mexican immigration because it increases U.S. economic activity, which, in turn, attracts new businesses and creates more jobs for all residents. However, some argue that the incoming Mexican population puts pressures on individual states by having to provide immigrants with benefits for which they do not pay taxes. Regardless, the debate on the impact that the increasing number of Mexican immigrants has on the United States is still ongoing.

The second part of this monograph describes Mexico's economic and social climate, two of the most-important catalysts of Mexican immigration to the United States. Although stable compared with its last major crisis, in 1994–1995, Mexico's economic and social situation is in need of vast reform in many areas. Our examination of the factors yielding an increase in migration and the state of the Mexican economy suggests that poverty in Mexico is still very widespread and that, although economic inequality has not grown since 1996, neither has it been reduced.

Despite the intended benefits for Mexico with the establishment of NAFTA, which created the world's largest free-trade area and several economic reforms of the 1990s, Mexico's economic competitiveness with other countries still remains weak. The country lags behind most other developing countries in terms of competitiveness rankings; for 2011, the World Bank and the World Economic Forum rank Mexico 35th and 58th, respectively (see Chapter Nine). Although President Calderón has passed legislation to reform the country's weak judicial system and improve a generally excessive bureaucracy, these factors are generally thought to feed an underdeveloped competition culture. It has been estimated that regulatory burden alone costs Mexico at least 15 percent of its GDP.

Fiscal policy in Mexico faces enormous challenges as well. Comparison with other OECD countries indicates that Mexico has one of the lowest tax-to-GDP ratios. Moreover, the government remains dependent on the revenue of PEMEX, Mexico's state-owned oil monopoly. More than 30 percent of government revenue stems from oil taxes; although Mexico was the seventh-largest oil producer in the world as of 2010, production at the primary oil fields is projected to decline rapidly in the next decade. PEMEX has just begun to invest in the kind of R&D needed to compete in the global market, but many still believe that the failing oil reserves will further harm Mexico's fiscal well-being. Finally, there is increasing evidence of widespread tax evasion by Mexico's large informal sector of workers, which is estimated to account for 20 percent of the profits generated in the country (see Chapter Nine).

Much of Mexico's trade is with the United States. Mexico has become the third-largest trading partner of the United States during the post-NAFTA years, following only Canada and China. One of the objectives of NAFTA was that greater international economic integration of Canada, Mexico, and the United States would fuel large inflows of FDI in the manufacturing sector by creating high-productivity manufacturing jobs. This would enhance prosperity in all regions and alleviate some of Mexico's fiscal deficit. However, Mexico's share of U.S. FDI outflows amounted to only 3.3 percent in 1993, 3.0 percent in 2009, and only 0.6 percent in 2010 (see Chapter Nine).

Although we found that there is widespread poverty in Mexico, the country also has a long tradition of instituting social programs with a primary goal of alleviating this poverty. The largest program, Oportunidades, has helped many citizens by providing cash transfers to households in the most-disadvantaged segments of society, but there are still many day-to-day challenges for individuals and families facing the stresses of making ends meet. More needs to be done in this area to eliminate inefficient social programs and target effectively those in most need.

By all accounts, the overall quality of education in Mexico is very low. The country has made impressive gains in increasing access to basic learning through ambitious reforms proposed by President Calderón. However, poor scores on academic achievement tests and a high dropout rate among both the urban and rural populations suggest that there is still much work to be done in this area. Almost 80 percent of the education budget is given to teacher compensation. This leaves little left to invest in other resources.

Residents' well-being and quality of life depends on their good health as well as sufficient health care, and, in these areas, Mexico has made great strides in the past century and particularly in the past decade. Life expectancy has increased by almost 28 years between 1950 and 2010, and infant mortality declined more than 60 percent between 1990 and 2010. Although rural populations remain at risk of death due to transmissible disease and malnutrition, sources reveal that Mexico overall is undergoing an epidemiological transition—a change in cause-of-death patterns. This shift suggests that the functional health status of the population is improving as more people live longer and develop chronic degenerative diseases characteristic of old age.

One cause of concern is the relatively low level of resources that the government allocates to the health system. In the past decade, one of the main government initiatives is Seguro Popular. Seguro Popular is a government program that specifically targets the poor and workers in the informal sector and has expanded its coverage substantially. As the health system stands, health coverage in Mexico is still highly fragmented; although changes to the country's general health law attempt to increase health spending and efficiency, more work remains to be done in this area.

Although there have been calls for deeper policy integration between the United States and Mexico, there was not much progress during the 2000s. Immigration proposals during the 2000s were halted in Congress, and safety and environmental concerns created disincentives to operate trucking according to NAFTA promises of accelerated FDI. Public opinion polls suggest that, although most U.S. citizens want the number of illegal Mexican immigrants to decrease in the future, many recognize the contributions Mexican-born U.S. citizens offer their new country.

APPENDIX

Political Contexts Behind Mexican Reforms

A brief summary of the political context behind Mexico's economic and social reforms will provide understanding for the demographic, economic, and social resources on which reforms and current policy are founded. The key points include democratization, competitive electoral races for the presidency, and the rising assertiveness of Congress.

Starting in 1929, the PRI established itself as the ruler of a one-party state, with society organized along corporatist lines. Its preeminence began to break down after the massacre of possibly thousands of students at a mass protest outside the foreign ministry in the Tlatelolco neighborhood of Mexico City in 1968, and economic crises in 1982 and 1994 resulted in further reforms (Preston and Dillon, 2004).

Constitutional reforms in 1996 improved the autonomy of the Federal Electoral Institute (Instituto Federal Electoral, or IFE), which oversees elections throughout the republic, and gave the federal judiciary jurisdiction over electoral appeals of local disputes (Eisenstadt, 2007). In 1997, the PRI lost its majority in the Chamber of Deputies for the first time, and, in 2000, the PRI lost the presidency for the first time. Mexico's 2000 presidential election, widely accepted as fair, resulted in the election of Vicente Fox of the PAN and the end of the 71-year rule of the PRI. For the first time since 1970, a financial crisis did not occur in the presidential election year. And the transfer of power took place smoothly when Fox was sworn in as president on December 1, 2000.

In the election of 2006, PAN candidate Felipe Calderón achieved a narrow victory after trailing Andrés Manuel López Obrador, the candidate of the PRD. López Obrador refused to concede, initiating mass protests and shutting down parts of Mexico City, despite Mexico's independent election institutions vetting the election's fairness. On December 1, Calderón was inaugurated president in a tightly guarded ceremony that lacked the celebration of previous inaugurations. The other noteworthy fact about the 2006 election, besides the narrow margin between the two front runners, was the absolute drubbing taken by the PRI candidate, Roberto Madrazo, who received less than one-quarter of the votes. Mexico will have its next presidential election in 2012, and there will be a new president elected because Mexican presidents can hold only one term.

Along with these changes, the Congress, which includes a Chamber of Deputies and a Senate, has become more assertive and a far more important player in Mexican government since the democratization measures of the 1990s. Indeed, congressional stalemate from 2003 to 2006 blocked Fox's reform program (Langston, 2007). Congresses are numbered in three-year terms; deputies have three-year terms and senators have six-year terms. In the 2006–2009 Congress, once again, no party had a majority, so that legislative reforms, which need majority vote, required compromise between at least two of the major parties. The PRI made dramatic gains in the 2009–2012 Congress, ending up with 49 percent of the seats in the Chamber of Deputies and an absolute majority in an alliance with the small Green Party (Selee and Putnam, 2009). However, the composition of the Senate did not change.

Constitutional reforms, viewed as necessary to continue modernizing Mexico and improving its economy, were certainly difficult to pass in the 2006–2009 Congress, and passage remained difficult in the 2009–2012 Congress. These require a two-thirds vote in both houses and two-thirds of the state legislatures. In neither house do any two major parties have two-thirds of the seats. However, including election coalition partners, the PAN and the PRI actually have two-thirds in both houses in the 2009–2012 Congress.

References

Adams, John A., *Bordering the Future: The Impact of Mexico on the United States*, Westport, Conn.: Praeger, 2006.

Adams, Richard H., Jr., and John Page, "Do International Migration and Remittances Reduce Poverty in Developing Countries?" in Organisation for Economic Co-Operation and Development, *Migration, Remittances and Development*, Paris, 2005, pp. 217–246.

Aguila, Emma, "Personal Retirement Accounts and Saving," *American Economic Journal: Economic Policy*, Vol. 3, No. 4, November 2011, pp. 1–24.

Aguila, Emma, Nelly Aguilera, and Cesar Velázquez, "Impact of Pension Reform on the Labor Market in Mexico: Evidence Using Administrative Data," Inter-American Conference on Social Security, 2008.

Aguila, Emma, Claudia Diaz, Mary Manqing Fu, Arie Kapteyn, and Ashley Pierson, *Living Longer in Mexico: Income Security and Health*, Santa Monica, Calif.: RAND Corporation, MG-1179-CF/AARP, 2011. As of January 30, 2012:
http://www.rand.org/pubs/monographs/MG1179.html

Aguila, Emma, and Julie Zissimopoulos, *Labor Market and Immigration Behavior of Middle-Aged and Elderly Mexicans*, Santa Monica, Calif.: RAND Corporation, WR-726, 2010. As of July 11, 2011:
http://www.rand.org/pubs/working_papers/WR726.html

Ahmad, Ehtisham, José Antonio González Anaya, Giorgio Brosio, Mercedes García-Escribano, Ben Lockwood, and Ernesto Revilla, *Why Focus on Spending Needs Factors? The Political Economy of Fiscal Transfer Reforms in Mexico*, Washington, D.C.: International Monetary Fund, Fiscal Affairs Department, October 2007. As of January 30, 2012:
http://www.imf.org/external/pubs/cat/longres.cfm?sk=21381.0

Alarcón, Rafael, "El retorno de los migrantes Mexicanos," *La jornada*, October 28, 2008. As of August 9, 2011:
http://www.jornada.unam.mx/2008/10/28/index.php?section=opinion&article=016a1pol

Alvarado Nieto, Jesús, and Luis C. Carbajo, "Americans Say Free Trade Good for Economy but Costs Jobs," *Rasmussen Reports*, November 12, 2010. As of August 9, 2011:
http://www.rasmussenreports.com/public_content/business/general_business/november_2010/americans_say_free_trade_good_for_economy_but_costs_jobs

Amuedo-Dorantes, Catalina, and Kusum Mundra, "Social Networks and Their Impact on the Earnings of Mexican Migrants," *Demography*, Vol. 44, No. 4, November 2007, pp. 849–863.

Amuedo-Dorantes, Catalina, Tania Sainz, and Susan Pozo, *Remittances and Healthcare Expenditure Patterns of Populations in Origin Communities: Evidence from Mexico*, Washington, D.C.: Inter-American Development Bank, Institute for the Integration of Latin America and the Caribbean/Integration, Trade and Hemispheric Issues Division, working paper 25, February 2007. As of April 29, 2010:
http://www.iadb.org/intal/aplicaciones/uploads/publicaciones/i_INTALITD_WP_25_2007_AmuedoDorantes_Sainz_Pozo.pdf

Andere, Eduardo, *La educación en México: Un fracaso monumental—está México en riesgo?* México: Planeta, 2003.

Angelucci, Manuela, *Aid Programs'* Unintended *Effects: The Case of Progresa and Migration*, Tucson, Ariz.: University of Arizona, Department of Economics, July 2005. As of April 29, 2010:
http://econ.arizona.edu/docs/Working_Papers/Econ-WP-05-16.pdf

Antón, Arturo, Fausto Hernández, and Santiago Levy, *The End of Informality in Mexico? Fiscal Reform for Universal Social Insurance*, Inter-American Development Bank, 2011.

Aspe Armella, Pedro, *El camino Mexicano de la transformación económica*, 2nd ed. in Spanish, México: Fondo de Cultura Económica, 1993.

Assistant Secretary for Planning and Evaluation, U.S. Department of Health and Human Services, "Summary of Immigrant Eligibility Restrictions Under Current Law as of 2/25/2009," last updated April 28, 2011. As of August 9, 2011:
http://aspe.hhs.gov/hsp/immigration/restrictions-sum.shtml

Aydemir, Abdurrahman, and George J. Borjas, *A Comparative Analysis of the Labor Market Impact of International Migration: Canada, Mexico, and the United States*, Cambridge, Mass.: National Bureau of Economic Research, working paper 12327, June 2006. As of August 9, 2011:
http://www.nber.org/papers/w12327

Baker and Associates, *Oil and Gas Policies in Mexico*, May 16, 2006.

Banco de México, "Estadísticas," undated web page, referenced December 14, 2008. As of January 30, 2012:
http://www.banxico.org.mx/estadisticas/index.html

———, *Addendum to the Inflation Report July–September 2009*, quarterly inflation report, December 2, 2009. As of October 5, 2010:
http://www.banxico.org.mx/publicaciones-y-discursos/publicaciones/informes-periodicos/trimestral-inflacion/%7BC915D461-0A6C-70CE-DCB1-425E415B0979%7D.pdf

BANXICO—*See* Banco de México.

Barefoot, Kevin B., and Marilyn Ibarra-Caton, "Direct Investment Positions for 2010: Country and Industry Detail," *Survey of Current Business*, July 2011, pp. 125–141. As of January 30, 2012:
http://www.bea.gov/scb/pdf/2011/07%20July/0711_direct.pdf

BEA—*See* Bureau of Economic Analysis.

Bergin, Paul R., Robert C. Feenstra, and Gordon H. Hanson, "Offshoring and Volatility: Evidence from Mexico's Maquiladora Industry," *American Economic Review*, Vol. 99, No. 4, September 2009, pp. 1664–1671.

Binational Study on Migration, *Migration Between Mexico and the United States: Binational Study*, Washington, D.C.: Commission on Immigration Reform, 1997.

Bird, Richard M., and Michael Smart, *Intergovernmental Fiscal Transfers: Some Lessons from International Experience*, Toronto: University of Toronto Press, March 2001.

Board of Governors of the Federal Reserve System, "Data Download Program," undated. As of August 11, 2011:
http://www.federalreserve.gov/datadownload/Choose.aspx?rel=H15

Borjas, George J., "The Labor Demand Curve *Is* Downward Sloping: Reexamining the Impact of Immigration on the Labor Market," *Quarterly Journal of Economics*, Vol. 118, No. 4, November 2003, pp. 1335–1374.

———, "Native Internal Migration and the Labor Market Impact of Immigration," *Journal of Human Resources*, Vol. 41, No. 2, Spring 2006, pp. 221–258.

Botero, Juan, Simeon Djankov, Rafael La Porta, Florencio Lopez-de-Silanes, and Andrei Shliefer, *The Regulation of Labor*, Cambridge, Mass.: National Bureau of Economic Research, working paper 9756, June 2003. As of April 29, 2010:
http://www.nber.org/papers/W9756.pdf

BP, *Statistical Review of World Energy*, June 2008.

———, *Statistical Review of World Energy*, June 2009. As of January 30, 2012:
http://www.bp.com/liveassets/bp_internet/globalbp/globalbp_uk_english/reports_and_publications/statistical_energy_review_2008/STAGING/local_assets/2009_downloads/statistical_review_of_world_energy_full_report_2009.pdf

———, *Statistical Review of World Energy*, June 2011. As of January 30, 2012: http://www.bp.com/assets/bp_internet/globalbp/globalbp_uk_english/reports_and_publications/statistical_energy_review_2011/STAGING/local_assets/pdf/statistical_review_of_world_energy_full_report_2011.pdf

Brauer, Gustavo, "Maquila in Mexico," Berg and Duffy, undated web page. As of June 1, 2010:
http://www.bergduffy.com/Mexico/maquila_in_mexico.htm

Bremer, Catherine, "Analysis: Pemex Tax Break to Cushion Mexico's Oil Woes," Reuters, September 14, 2007. As of January 30, 2012:
http://uk.reuters.com/article/2007/09/14/mexico-oil-idUKN1422012220070914

Bruhn, Kathleen, and Kenneth F. Greene, "Elite Polarization Meets Mass Moderation in Mexico's 2006 Elections," *Political Science and Politics*, Vol. 40, No. 1, January 2007, pp. 33–38.

Bruno, Andorra, *Unauthorized Alien Students: Issues and "DREAM Act" Legislation*, Washington, D.C.: Congressional Research Service, CRS Report for Congress RL33863, March 22, 2011.

Bugarin, Alicia, *Inventory of Mexico Related Projects Conducted by California State Agencies*, Sacramento, Calif.: California State Library, California Research Bureau, report CRB 04-0011, November 2004. As of April 29, 2010:
http://bibpurl.oclc.org/web/9377

Bureau of Economic Analysis, balance-of-payments and direct investment position data, 2007a.

———, "Table D. Cross-Border Services Exports and Imports by Type and Country, 2005," June 2007b. As of June 1, 2010:
http://www.bea.gov/international/xls/tabD.xls

———, "U.S. International Trade in Goods and Services: Annual Revision for 2007," press release, June 10, 2008. As of May 4, 2010:
http://www.bea.gov/newsreleases/international/trade/2008/trad1308.htm

———, "National Economic Accounts: All NIPA Tables," last modified October 21, 2009.

———, "U.S. Direct Investment Abroad: Balance of Payments and Direct Investment Position Data," last modified March 18, 2010. As of April 29, 2010:
http://www.bea.gov/international/di1usdbal.htm

Burnside, Craig, and Yuliya Meshcheryakova, "Mexico: A Case Study of Procyclical Fiscal Policy," in Craig Burnside, ed., *Fiscal Sustainability in Theory and Practice: A Handbook*, Washington, D.C.: World Bank, 2005, pp. 133–174.

Bustamante, Jorge A., Guillermina Jasso, J. Edward Taylor, and Paz Trigueros Legarreta, "Characteristics of Migrants: Mexicans in the United States," in *Migration Between Mexico and the United States: Binational Study*, Vol. 1: *Thematic Chapters*, Mexico–United States Binational Migration Study, 1998, pp. 91–162. As of April 29, 2010:
http://www.utexas.edu/lbj/uscir/binpap-v.html

Cámara de Diputados del H. Congreso de la Unión, Ley general de salud, April 27, 2010. As of May 4, 2010:
http://www.diputados.gob.mx/LeyesBiblio/pdf/142.pdf

Camarena, Rodrigo, "Mexico's Energy Reform and the Future of Pemex," *World Politics Review*, October 19, 2010. As of November 1, 2010:
http://www.worldpoliticsreview.com/articles/6759/mexicos-energy-reform-and-the-future-of-pemex

Camarota, Steven A., and Karen Jensenius, *A Shifting Tide: Recent Trends in the Illegal Immigrant Population*, Center for Immigration Studies, backgrounder, July 2009. As of August 9, 2011:
http://www.cis.org/illegalimmigration-shiftingtide

Campbell, Robert, "Update 2: Mexico New Oil Reserves Outweigh Output in 2008," Reuters, March 18, 2009. As of January 30, 2012:
http://www.reuters.com/article/2009/03/18/mexico-oil-idUSN1840014820090318

Canales, Alejandro, "Las remesas de los migrantes: Fondos para el ahorro o ingresos salariales?" in Germán A. Zárate Hoyos, ed., *Remesas de los mexicanos y centroamericanos en Estados Unidos: problemas y perspectivas*, México: Miguel Ángel Porrúa, El Colegio de La Frontera Norte, 2004, pp. 97–128.

Cañas, Jesus, and Robert W. Gilmer, "Spotlight: Maquiladora Data—Mexican Reform Clouds View of Key Industry," *Southwest Economy*, Vol. 3, May–June 2007. As of August 9, 2011:
http://dallasfed.org/research/swe/2007/swe0703d.cfm

Capps, Randy, Michael Fix, and Serena Yi-Ying Lin, *Still an Hourglass? Immigrant Workers in Middle-Skilled Jobs*, Migration Policy Institute, September 2010. As of August 9, 2011:
http://www.migrationpolicy.org/pubs/sectoralstudy-Sept2010.pdf

Capps, Randolph, Karina Fortuny, and Michael E. Fix, *Trends in the Low-Wage Immigrant Labor Force, 2000–2005*, Washington, D.C.: Urban Institute, March 6, 2007. As of January 30, 2012:
http://www.urban.org/publications/411426.html

Card, David E., "Immigrant Inflows, Native Outflows, and the Local Labor Market Impacts of Higher Immigration," *Journal of Labor Economics*, Vol. 19, No. 1, January 2001, pp. 22–64.

———, "How Immigration Affects U.S. Cities," London: Centre for Research and Analysis of Migration, discussion paper 11/07, June 2007.

———, *Immigration and Inequality*, Cambridge, Mass.: National Bureau of Economic Research, working paper 14683, January 2009. As of October 5, 2010: http://www.nber.org/papers/w14683

Card, David E., and John E. DiNardo, *Do Immigrant Inflows Lead to Native Outflows?* Cambridge, Mass.: National Bureau of Economic Research, working paper 7578, March 2000. As of February 16, 2012: http://www.nber.org/papers/w7578

Card, David E., and Ethan G. Lewis, *The Diffusion of Mexican Immigrants During the 1990s: Explanations and Impacts*, Cambridge, Mass.: National Bureau of Economic Research, working paper 11552, August 2005. As of January 30, 2012: http://www.nber.org/papers/w11552

Castañeda, Jorge G., *Primer informe de labores del Secretario de Relaciones Exteriores*, Mexico City, México, June 14, 2001.

———, "The Challenge of Development in the U.S.-Mexican Context," *Trans-Border Migration and Development Conference: Promoting Economic Opportunities in Mexico and the Border Region*, October 5, 2006.

CBS News/New York Times, poll, May 18–23, 2007, N = 1,125 adults nationwide.

CDI—*See* Comisión Nacional para el Desarrollo de los Pueblos Indígenas.

CEESP—*See* Centro de Estudios Económicos del Sector Privado.

Centro de Estudios Económicos del Sector Privado, *Encuesta sobre gobernabilidad y desarrollo empresarial 2005*, August 2005. As of April 29, 2010: http://www.contraloriaciudadana.org.mx/mediciones/nacionales/encuestasobregobernabilidad/Prensa3edge170805.pdf

Centro de Estudios de las Finanzas Publicas, *El ingreso tributario en Mexico*, 2005.

Chamberlain, Andrew, "Corporate Income Tax Rates Around the World," *Tax Foundation Tax Policy Blog*, May 5, 2006. As of January 30, 2012: http://www.taxfoundation.org/blog/show/1471.html

Chiquiar, Daniel, and Gordon H. Hanson, "International Migration, Self-Selection, and the Distribution of Wages: Evidence from Mexico and the United States," *Journal of Political Economy*, Vol. 113, No. 2, April 2005, pp. 239–281.

Chiquiar, Daniel, and Manuel Ramos-Francia, *Competitiveness and Growth of the Mexican Economy*, Banco de México Documentos de Investigación 2009-11, November 2009. As of August 9, 2011: http://www.banxico.org.mx/publicaciones-y-discursos/publicaciones/documentos-de-investigacion/banxico/%7BA07F4B14-3BC3-2B7C-361B-97DA7C4296CF%7D.pdf

CNN/Opinion Research Corporation, poll, January 14–17, 2008, N = 1,038 adults nationwide.

COLEF—*See* Colegio de la Frontera Norte.

Colegio de la Frontera Norte, "Encuestas sobre Migración en las Fronteras Norte y Sur de México," undated web page. As of June 1, 2010:
http://www.colef.mx/emif/

Coleman, David, "Immigration and Ethnic Change in Low-Fertility Countries: A Third Demographic Transition," *Population and Development Review*, Vol. 32, No. 3, September 2006, pp. 401–446.

Collier, Robert, "Mexico's Trucks on Horizon: Long-Distance Haulers Are Headed into U.S. Once Bush Opens Borders," *San Francisco Chronicle*, March 4, 2001. As of January 30, 2012:
http://www.sfgate.com/cgi-bin/article.cgi?f=/c/a/2001/03/04/MN143818.DTL

Comisión Nacional para el Desarrollo de los Pueblos Indígenas, *Indicadores sociodemográficos de la población indígena 2000–2005*, September 2006.

Comisión Nacional de los Salarios Mínimos, "Salarios mínimos," last modified December 21, 2009.

CONAPO—*See* Consejo Nacional de Población.

CONASAMI—*See* Comisión Nacional de los Salarios Mínimos.

CONEVAL—*See* Consejo Nacional de Evaluación de la Política de Desarrollo Social.

Consejo Nacional de Evaluación de la Política de Desarrollo Social, "Pobreza por ingresos," undated (a).

———, "Evolución de las dimensiones de la pobreza 1990–2010: Anexo," undated (b), referenced September 12, 2011. As of January 30, 2012:
http://www.coneval.gob.mx/cmsconeval/rw/pages/medicion/evolucion_de_las_dimensiones_pobreza_1990_2010.en.do

———, "El CONEVAL reporta cifras Sobre la Evolución de la Pobreza en México," October 1, 2006.

———, "Reporta CONEVAL cifras de pobreza por ingresos 2008," Distrito Federal, Comunicado de Prensa 006/09, July 18, 2009. As of August 9, 2011:
http://www.coneval.gob.mx/contenido/home/3491.pdf

Consejo Nacional de Población, *2005, Municipales y estatales*, c. 2005a.

———, *Proyecciones de indígenas de México y de las entidades federativas 2000–2010*, México, 2005b.

———, *Proyecciones de población de México, 2005–2050*, México, 2006.

———, "Indicadores demográficos básicos 1990–2050," 2007.

Cornelius, Wayne A., and Idean Salehyan, "Does Border Enforcement Deter Unauthorized Immigration? The Case of Mexican Migration to the United States of America," *Regulation and Governance*, Vol. 1, 2007, pp. 139–153.

Cortes, Patricia, "The Effect of Low-Skilled Immigration on U.S. Prices: Evidence from CPI Data," *Journal of Political Economy*, Vol. 116, No. 3, June 2008, pp. 381–422.

Cotis, Jean-Philippe, *What Are the OECD's Views About the Mexican Tax Reform*, October 13–14, 2003. As of April 29, 2010:
http://www.oecd.org/LongAbstract/0,3425,en_33873108_33873610_22425200_1_1_1_1,00.html

Council of Economic Advisers, *Immigration's Economic Impact*, Washington, D.C.: Executive Office of the President, Council of Economic Advisers, June 20, 2007.

Crain, W. Mark, and Thomas D. Hopkins, *The Impact of Regulatory Costs on Small Firms*, Washington, D.C.: U.S. Small Business Administration, Office of Advocacy, 2001. As of June 1, 2010:
http://purl.access.gpo.gov/GPO/LPS95322

Creighton, Mathew J., Hyunjoon Park, and Graciela M. Teruel, "The Role of Migration and Single Motherhood in Upper Secondary Education in Mexico," *Journal of Marriage and Family*, Vol. 71, No. 5, 2009, pp. 1325–1339.

Day, Jennifer Cheeseman, *Population Projections of the United States, by Age, Sex, Race, and Hispanic Origin: 1993 to 2050*, U.S. Census Bureau, Current Population Reports, P25-1104, November 1993. As of August 9, 2011:
http://www.census.gov/prod/1/pop/p25-1104.pdf

De la Torre, Rodolfo, and A. Héctor Moreno, "La distribución de la riqueza y del ingreso factorial en México," in García Alba Iduñate, Lucino Gutiérrez Herrera Pascual, and Gabriela Torres Ramírez, eds., *El nuevo milenio Mexicano*, Tomo 4: *Los retos sociales*, Universidad Autónoma Metropolitana, Unidad Azcapotzalco: Ediciones y Gráficos Eón, 2004.

DeLaet, Debra L., *U.S. Immigration Policy in an Age of Rights*, Westport, Conn.: Praeger, 2000.

Deloitte Touche Tohmatsu, *International Tax and Business Guide: Mexico*, August 2005.

Diaz, Lizbeth, "Drug War Shutters Businesses on Mexico Border," Reuters, May 15, 2008. As of January 30, 2012:
http://www.reuters.com/article/2008/05/15/us-mexico-tijuana-idUSN0531742620080515

Diaz-Cayeros, Alberto, "Dependencia fiscal y estrategias de coalición en el federalismo Mexicano," *Política y gobierno,* Vol. XI, No. 2, 2004, pp. 229–262. As of January 30, 2012:
http://www.politicaygobierno.cide.edu/num_anteriores/Vol_XI_N2_2004/DIAZ-CAYEROS.pdf

Diaz-Cayeros, Alberto, José Antonio González, and Fernando Rojas, *Mexico's Decentralization at a Cross-Roads*, Stanford University Center for Research on Economic Development and Policy Reform, working paper, 2002.

Dinan, Stephen, "Immigration Bill Quashed; Senators Swayed by Pressure from Public," *Washington Times*, June 29, 2007.

Dinan, Stephen, and Sean Lengell, "Immigration Deal Called a 'Sellout'; Democrats Warned by Advocates," *Washington Times*, May 10, 2007.

Donato, Katharine M., Jorge Durand, and Douglas S. Massey, "Changing Conditions in the US Labor Market," *Population Research and Policy Review*, Vol. 11, No. 2, January 1992, pp. 93–115.

Economic Research Service, U.S. Department of Agriculture, "Transportation Bottlenecks Shape U.S.-Mexico Food and Agricultural Trade," *Agricultural Outlook*, September 2000. As of January 30, 2012:
http://www.ers.usda.gov/publications/agoutlook/sep2000/ao274h.pdf

Economist Intelligence Unit, "Mexico: Economic Data," February 19, 2008.

EIA—*See* Energy Information Administration.

Eisenstadt, Todd A., "The Origins and Rationality of the 'Legal Versus Legitimate' Dichotomy Invoked in Mexico's 2006 Post-Electoral Conflict," *PS: Political Science and Politics*, Vol. 40, No. 1, January 2007, pp. 39–43.

EIU—*See* Economist Intelligence Unit.

Embassy of Mexico, "Mexico and the Migration Phenomenon," March 2006. As of May 6, 2010:
http://www.ime.gob.mx/agenda_migratoria/congreso/mexico_and_the_migration_phenomenon.pdf

Energy Information Administration, "Country Analysis Briefs: Mexico," Washington, D.C.: Energy Information Administration, U.S. Department of Energy, December 2007a.

———, *Short-Term Energy Outlook*, Washington, D.C., 2007b.

———, "Mexico," *Country Analysis Briefs,* U.S. Department of Energy, July 2011.

English, Cynthia, "Opinion Briefing: North American Free Trade Agreement," Gallup, December 12, 2008. As of August 9, 2011:
http://www.gallup.com/poll/113200/Opinion-Briefing-North-American-Free-Trade-Agreement.aspx

Environmental Systems Research Institute, *esri Data and Maps*, version 10, Redlands, Calif., undated.

Equivel, Gerado, *The Dynamics of Income Inequality in Mexico Since NAFTA*, Colegio de México, working paper No. IX-2010, 2010.

ERS—*See* Economic Research Service/USDA.

Escobar Latapí, Agustín, and Susan Forbes Martin, eds., *La gestión de la migración México–Estados Unidos: Un enfoque binacional*, México: SEGOB, Instituto Nacional de Migración, Centro de Estudios Migratorios, 2008.

Fernández de Castro, Rafael, Rodolfo García Zamora, Roberta Clariond Rangel, and Ana Vila Freyer, eds., *Las políticas migratorias en los estados de México: Una evaluación*, México, D.F.: Cámara de Diputados, LX Legislatura, 2007.

Figueroa-Aramoni, Rodulfo, "A Nation Beyond Its Borders: The Program for Mexican Communities Abroad," *Journal of American History*, Vol. 86, No. 2, September 1999, pp. 537–544. As of August 9, 2011:
http://www.journalofamericanhistory.org/projects/mexico/rfigueroa.html

Fitzgerald, David, "Inside the Sending State: The Politics of Mexican Emigration Control," *International Migration Review*, Vol. 40, No. 2, June 2006, pp. 259–293.

Fleck, Susan, and Constance Sorrentino, "Employment and Unemployment in Mexico's Labor Force," *Monthly Labor Review*, Vol. 117, No. 11, November 1994.

Fox, Vicente, president of the United Mexican States, address to the U.S. House of Representatives, September 6, 2001.

Frey, William H., and Kao-Lee Liaw, "The Impact of Recent Immigration on Population Redistribution Within the United States," in James P. Smith and Barry Edmonston, eds., *The Immigration Debate: Studies on the Economic, Demographic, and Fiscal Effects of Immigration*, Washington, D.C.: National Academies Press, 1998, pp. 388–448. As of August 9, 2011:
http://www.nap.edu/catalog.php?record_id=5985

Friedberg, Rachel, and Jennifer Hunt, "The Economic Impact of Immigrants on Host Country Wages, Employment, and Growth," *Journal of Economic Perspectives*, Vol. 9, No. 2, Spring 1995, pp. 23–44.

Fundación Mexicana para la Salud, *La salud en México: 2006/2012*, México, D.F., 2006.

FUNSALUD—*See* Fundación Mexicana para la Salud.

Gallup, immigration poll, June 2008. As of December 14, 2008:
http://www.gallup.com/poll/1660/Immigration.aspx

García Zamora, Rodolfo, "Mexico: International Migration, Remittances and Development," in Organisation for Economic Co-Operation and Development, *Migration, Remittances and Development*, Paris, 2005a, pp. 81–87.

———, *Migración internacional y remesas: Las lecciones del Programa Tres por Uno en Zacatecas*, Zacatecas, México: Universidad Autónoma de Zacatecas, 2005b.

Gibson, Campbell J., and Emily Lennon, *Historical Census Statistics on the Foreign-Born Population of the United States: 1850–1990*, Washington, D.C.: Population Division, U.S. Census Bureau, working paper 29, February 1999. As of April 29, 2010:
http://www.census.gov/population/www/documentation/twps0029/twps0029.html

Giorguli Saucedo, Silvia E., Selene Gaspar Olvera, and Paula Leite, *La migración Mexicana y el mercado laboral estadounidense: Tendencias, perspectivas, y ¿oportunidades?* México, D.F.: Consejo Nacional de Población, 2006.

Gonzales, Felisa, "Statistical Portrait of the Foreign-Born Population in the United States, 2006," Pew Hispanic Center, January 23, 2008. As of January 31, 2012:
http://www.pewhispanic.org/2008/01/23/statistical-portrait-of-the-foreign-born-population-in-the-united-states-2006/

Gould, David M., "Immigrant Links to the Home Country: Empirical Implications for U.S. Bilateral Trade Flows," *Review of Economics and Statistics*, Vol. 76, No. 2, May 1994, pp. 302–316.

Grubel, Herbert G., and P. J. Lloyd, *Intra-Industry Trade: The Theory and Measurement of International Trade in Differentiated Products*, New York: Wiley, 1975.

Guarnizo, Luis Eduardo, "The Economics of Transnational Living," *International Migration Review*, Vol. 37, No. 3, Fall 2003, pp. 666–699.

Guichard, Stéphanie, *The Education Challenge in Mexico: Delivering Good Quality Education to All*, Paris: Organisation for Economic Co-Operation and Development, working paper 447, September 30, 2005.

Hanson, Gordon H., "Illegal Migration from Mexico to the United States," *Journal of Economic Literature*, Vol. 44, No. 4, December 2006, pp. 869–924.

———, *The Economic Logic of Illegal Immigration*, New York: Council on Foreign Relations, 2007.

———, *The Economics and Policy of Illegal Immigration in the United States*, Washington, D.C.: Migration Policy Institute, 2009.

———, "Why Isn't Mexico Rich?" *Journal of Economic Literature*, Vol. 48, No. 4, December 2010, pp. 987–1004.

Hanson, Gordon H., and Craig McIntosh, "The Demography of Mexican Migration to the United States," *American Economic Review: Papers and Proceedings 2009*, Vol. 99, No. 2, May 2009, pp. 22–27.

Hanson, Gordon H., and Christopher Woodruff, *Emigration and Educational Attainment in Mexico*, University of California at San Diego, unpublished, 2003.

Harrup, Anthony, "Mexico's Senate Approves Changes in Competition Law," *Wall Street Journal*, April 28, 2011. As of January 30, 2012:
http://online.wsj.com/article/
SB10001424052748704330404576291644040999126.html

Hertel, Thomas W., Roman Keeney, Maros Ivanic, and L. Alan Winters, *Why Isn't the Doha Development Agenda More Poverty Friendly?* Center for Global Trade Analysis, Department of Agricultural Economics, Purdue University, working paper 37, 2007. As of April 29, 2010:
https://www.gtap.agecon.purdue.edu/resources/res_display.asp?RecordID=2292

Hoefer, Michael, Nancy Rytina, and Bryan C. Baker, *Estimates of the Unauthorized Immigrant Population Residing in the United States: January 2008*, Washington, D.C.: Office of Immigration Statistics, Department of Homeland Security, February 2009.

Homedes, Núria, and Antonio Ugalde, "Decentralization of Health Services in Mexico: A Historical Review," in Núria Homedes and Antonio Ugalde, eds., *Decentralizing Health Services in Mexico: A Case Study in State Reform*, La Jolla, Calif.: Center for U.S.-Mexican Studies, 2006, pp. 45–94.

Hussein, Abdel Fattah, "Employment Grows in Mexico . . . Informal Reveals INEGI," *Allvoices*, June 13, 2011. As of November 1, 2011:
http://www.allvoices.com/contributed-news/
9380862-employment-grows-in-mexico-informal-reveals-inegi

IADB—*See* Inter-American Development Bank.

IMCO—*See* Instituto Mexicano para la Competitividad.

IMF—*See* International Monetary Fund.

INEGI—*See* Instituto Nacional de Estadística y Geografía.

Instituto Mexicano para la Competitividad, *Situación de la competitividad de México 2004: Hacia un pacto de competitividad*, México, 2005.

Instituto Nacional de Estadística y Geografía, undated (a) homepage, referenced December 2008. As of January 30, 2012:
http://www.inegi.org.mx/default.aspx

———, "Encuesta Nacional de Ocupación y Empleo (ENOE): Tabulados básicos—descarga," undated (b).

———, *Censos de población y vivienda*, 1950.

———, *Estadísticas del sector salud y seguridad social*, México, 2001.

———, *Censos de población y vivienda*, 2005a.

———, *Población rural y rural ampliada en México, 2000*, Aguascalientes, México, 2005b.

———, *II conteo de población y vivienda 2005*, Aguascalientes, México, 2006.

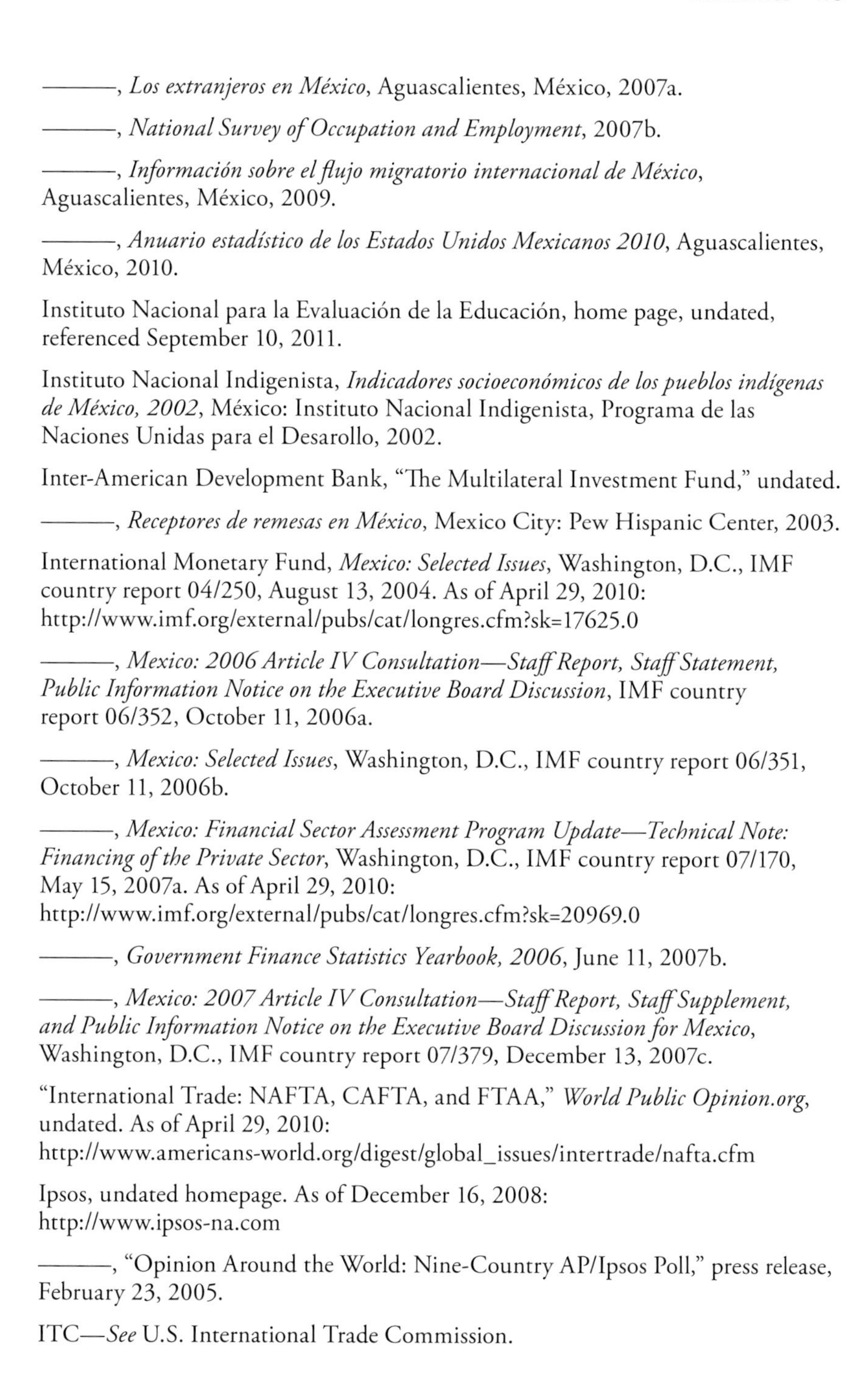

———, *Los extranjeros en México*, Aguascalientes, México, 2007a.

———, *National Survey of Occupation and Employment*, 2007b.

———, *Información sobre el flujo migratorio internacional de México*, Aguascalientes, México, 2009.

———, *Anuario estadístico de los Estados Unidos Mexicanos 2010*, Aguascalientes, México, 2010.

Instituto Nacional para la Evaluación de la Educación, home page, undated, referenced September 10, 2011.

Instituto Nacional Indigenista, *Indicadores socioeconómicos de los pueblos indígenas de México, 2002*, México: Instituto Nacional Indigenista, Programa de las Naciones Unidas para el Desarollo, 2002.

Inter-American Development Bank, "The Multilateral Investment Fund," undated.

———, *Receptores de remesas en México*, Mexico City: Pew Hispanic Center, 2003.

International Monetary Fund, *Mexico: Selected Issues*, Washington, D.C., IMF country report 04/250, August 13, 2004. As of April 29, 2010: http://www.imf.org/external/pubs/cat/longres.cfm?sk=17625.0

———, *Mexico: 2006 Article IV Consultation—Staff Report, Staff Statement, Public Information Notice on the Executive Board Discussion*, IMF country report 06/352, October 11, 2006a.

———, *Mexico: Selected Issues*, Washington, D.C., IMF country report 06/351, October 11, 2006b.

———, *Mexico: Financial Sector Assessment Program Update—Technical Note: Financing of the Private Sector*, Washington, D.C., IMF country report 07/170, May 15, 2007a. As of April 29, 2010: http://www.imf.org/external/pubs/cat/longres.cfm?sk=20969.0

———, *Government Finance Statistics Yearbook, 2006*, June 11, 2007b.

———, *Mexico: 2007 Article IV Consultation—Staff Report, Staff Supplement, and Public Information Notice on the Executive Board Discussion for Mexico*, Washington, D.C., IMF country report 07/379, December 13, 2007c.

"International Trade: NAFTA, CAFTA, and FTAA," *World Public Opinion.org*, undated. As of April 29, 2010: http://www.americans-world.org/digest/global_issues/intertrade/nafta.cfm

Ipsos, undated homepage. As of December 16, 2008: http://www.ipsos-na.com

———, "Opinion Around the World: Nine-Country AP/Ipsos Poll," press release, February 23, 2005.

ITC—*See* U.S. International Trade Commission.

Itzigsohn, José, and Silvia Giorguli Saucedo, "Immigrant Incorporation and Sociocultural Transnationalism," *International Migration Review*, Vol. 36, No. 3, Fall 2002, pp. 766–798.

Jones, Jeffrey M., "Fewer Americans Favor Cutting Back Immigration: Public as Likely to Favor Status Quo as to Favor Decreased Immigration," Princeton, N.J.: Gallup, July 10, 2008. As of June 1, 2010:
http://www.gallup.com/poll/108748/
Fewer-Americans-Favor-Cutting-Back-Immigration.aspx

Jones, Laura, and Stephen Graf, "Canada's Regulatory Burden: How Many Regulations? At What Cost?" *Fraser Forum*, August 1, 2001.

Juárez Strategic Plan Association, *Juárez Strategic Plan: A Proposal Toward the Juárez We Envision*, Ciudad Juárez, Chihuahua, México, 2006. As of April 29, 2010:
http://www.planjuarez.org/files/pdf_168.pdf

Kehoe, Timothy J., and Kim J. Ruhl, "Why Have Economic Reforms in Mexico Not Generated Growth?" *Journal of Economic Literature*, Vol. 48, No. 4, December 2010, pp. 1005–1027.

Kellam, Beverly Fox, and Lucinda Vargas, "Immigration and the Economy, Part I," *Southwest Economy*, No. 4, July–August 1998. As of June 1, 2010:
http://www.dallasfed.org/research/swe/1998/swe9804.pdf

Koncz-Bruner, Jennifer, and Anne Flatness, "U.S. International Services: Cross-Border Trade in 2009 and Services Supplied Through Affiliates in 2008," *Survey of Current Business*, October 2010, pp. 18–60. As of August 9, 2011:
http://www.bea.gov/scb/pdf/2010/10%20October/1010_services.pdf

———, "U.S. International Services: Cross-Border Trade in 2010 and Services Supplied Through Affiliates in 2009," *Survey of Current Business*, October 2011, pp. 13–19. As of February 16, 2012:
http://www.bea.gov/scb/pdf/2011/10%20October/1011_services%20text.pdf

Kritz, Mary M., and Douglas T. Gurak, "The Impact of Immigration on the Internal Migration of Natives and Immigrants," *Demography*, Vol. 38, No. 1, 2001, pp. 133–145.

Langston, Joy, "The PRI's 2006 Electoral Debacle," *PS: Political Science and Politics*, Vol. 40, No. 1, January 2007, pp. 21–25.

LeMay, Michael C., *U.S. Immigration: A Reference Handbook*, Santa Barbara, Calif.: ABC-CLIO, 2004.

Levitt, Peggy, "Transnational Migrants: When 'Home' Means More Than One Country," *Migration Information Source*, October 2004. As of April 29, 2010:
http://www.migrationinformation.org/Feature/display.cfm?id=261

Levy, Santiago, *Good Intentions, Bad Outcomes: Social Policy, Informality, and Economic Growth in Mexico*, Washington, D.C.: Brookings Institution Press, 2008.

Lowe, Jeffrey H., "Direct Investment, 2007–2009: Detailed Historical-Cost Positions and Related Financial and Income Flows," *Survey of Current Business*, September 2010, pp. 53–58. As of August 9, 2011:
http://www.bea.gov/scb/pdf/2010/09%20September/0910_directinvest.pdf

Lund, Dan, "Migration Policies in Mexico, Beyond the Wall?" *Mund Américas Opinion Report*, Vol. 13, No. 6, October 6, 2006.

Lustig, Nora, *Mexico: The Remaking of an Economy*, 2nd ed., Washington, D.C.: Brookings Institution Press, 1998.

———, "Life Is Not Easy: Mexico's Quest for Stability and Growth," *Journal of Economic Perspectives*, Vol. 15, No. 1, Winter 2001, pp. 85–106.

Marrufo, Grecia, *The Incidence of Social Security Regulation in Economies with Partial Compliance: Evidence from the Reform in Mexico*, Chicago, Ill.: University of Chicago, 2001.

Martin, Philip, "Economic Integration and Migration: The Mexico–US Case," in United Nations University World Institute for Development Economics Research Conference on Poverty, *International Migration, and Asylum*, September 19, 2002.

———, *NAFTA and Mexico–US Migration: Policy Options in 2004*, presented at Institute for Research on Public Policy Conference on North American Integration, Montreal, Canada, April 1–2, 2004.

Martin, Susan, "The Politics of US Immigration Reform," in Sarah Spencer, ed., *The Politics of Migration: Managing Opportunity, Conflict and Change*, Malden, Mass.: Blackwell, 2003, pp. 132–149.

Massey, Douglas S., Jorge Durand, and Nolan J. Malone, *Beyond Smoke and Mirrors: Mexican Immigration in an Era of Economic Integration*, New York: Russell Sage Foundation, 2002.

Massey, Douglas S., and Felipe García España, "The Social Process of International Migration," *Science*, Vol. 237, No. 4816, August 14, 1987, pp. 733–738.

Massey, Douglas S., and Audrey Singer, "New Estimates of Undocumented Mexican Migration and the Probability of Apprehension," *Demography*, Vol. 32, No. 2, May 1995, pp. 203–213.

McDaniel, Christine A., and Laurie-Ann Agama, "The NAFTA Preference and U.S.-Mexico Trade: Aggregate-Level Analysis," *World Economy*, Vol. 26, No. 7, July 2003, pp. 939–955.

McGee, Marianne Kolbasuk, "With Immigration, Tech CEOs' Policy Pitch Runs into Washington Realities," *InformationWeek*, June 16, 2007. As of February 21, 2012:
http://www.informationweek.com/news/199904847

McKenzie, David J., and Hillel Rapoport, *Can Migration Reduce Educational Attainment? Evidence from Mexico*, Washington, D.C.: World Bank, Development Research Group, Growth and Investment Team, working paper 3952, June 2006.

———, "Network Effects and the Dynamics of Migration and Inequality: Theory and Evidence from Mexico," *Journal of Development Economics*, Vol. 84, No. 1, September 2007, pp. 1–24.

Medina Peña, Luis, *Hacia el nuevo estado: México, 1920–1994*, 2nd ed., México: Fondo de Cultura Económica, 1995.

Meissner, Doris, and Donald Kerwin, *DHS and Immigration: Taking Stock*, Migration Policy Institute, February 2009. As of August 9, 2011:
http://www.migrationpolicy.org/pubs/DHS_Feb09.pdf

Meissner, Doris, Deborah W. Meyers, Demetrios G. Papademetriou, and Michael Fix, *Immigration and America's Future: A New Chapter—Report of the Independent Task Force on Immigration and America's Future*, Washington, D.C.: Migration Policy Institute, 2006. As of April 29, 2010:
http://www.migrationpolicy.org/ITFIAF/finalreport.pdf

"Mexican Competition Law: Insight on the May 2011 Reform," Mexico City: White and Case, May 2011. As of January 31, 2012:
http://www.whitecase.com/alerts-05182011/

"Mexicans Divided on Proposed Energy Reform," *Angus Reid Public Opinion*, May 17, 2008. As of January 30, 2012:
http://www.angus-reid.com/polls/32030/
mexicans_divided_on_proposed_energy_reform/

"Mexico Hedges to Protect Oil Revenues," *Financial Times*, November 14, 2008.

"Mexico's 2010 Tax Reform Affects Individual and Corporate Taxes," *Maquila Watch*, Winter 2010. As of August 9, 2011:
http://www.napsintl.com/newsletter/MaqWatch09winterDec15.pdf

"Mexico's Declining Oil Industry," *Economist*, December 19, 2007.

"Mexico's Oil Dilemmas," *Economist*, February 24, 2005.

Migration Policy Institute, "US Historical Immigration Trends," undated web page. As of November 22, 2008:
http://www.migrationinformation.org/datahub/historicaltrends.cfm

———, *America's Emigrants: US Retirement Migration to Mexico and Panama*, Washington, D.C., 2006. As of April 29, 2010:
http://www.ngiweb.com/americasemigrants.pdf

Miller, Terry, Kim R. Holmes, and Edwin J. Feulner, with Anthony B. Kim, Bryan Riley, and James M. Roberts, *2012 Index of Economic Freedom*, Washington, D.C.: Heritage Foundation, November 2011. As of February 16, 2012:
http://www.heritage.org/index/

Mishra, Prachi, "Emigration and Wages in Source Countries: Evidence from Mexico," *Journal of Development Economics*, Vol. 82, No. 1, January 2007, pp. 180–199.

Moctezuma, Miguel, "Entusiasmo estatal por la inversión productiva de los Mexicanos que residen en el extranjero," in Carlos González Gutiérrez, ed., *Relaciones estado-diáspora: La perspectiva de América Latina y el Caribe*, México, D.F.: Miguel Ángel Porrúa, 2006, pp. 91–111.

Morales, Lymari, "Amid Immigration Debate, Americans' Views Ease Slightly," Gallup, July 27, 2010. As of August 9, 2011:
http://www.gallup.com/poll/141560/Amid-Immigration-Debate-Americans-Views-Ease-Slightly.aspx

Moreno, Carlos L., *Fiscal Performance of Local Governments in Mexico: The Role of Federal Transfers*, México: Centro de Investigación y Docencia Económicas, División de Administración Pública, 2003.

———, *Decentralization, Electoral Competition and Local Government Performance in Mexico*, Austin, Texas: University of Texas, 2005.

Munshi, Kaivan, "Networks in the Modern Economy: Mexican Migrants in the U.S. Labor Market," *Quarterly Journal of Economics*, Vol. 118, No. 2, May 2003, pp. 549–597.

Newport, Frank, "Economy Dominates as Nation's Most Important Problem," Gallup, July 14, 2010. As of August 9, 2011:
http://www.gallup.com/poll/141275/Economy-Dominates-Nation-Important-Problem.aspx

North American Free Trade Agreement Office of Mexico in Canada, "FTA's Signed by Mexico," Ottawa, Canada, undated.

North American Production Sharing, undated homepage. As of February 16, 2012:
http://www.napsintl.com/

"Obama Hopeful of Fixing Truck Dispute with Mexico," Reuters, April 16, 2009. As of January 31, 2012:
http://www.reuters.com/article/2009/04/16/idUSN16279139

OECD—*See* Organisation for Economic Co-Operation and Development.

OIG DoT—*See* U.S. Department of Transportation Office of Inspector General.

Organisation for Economic Co-Operation and Development, "Corporate Income Tax Rates," *OECD Tax Database*, Table II.1, undated (a), referenced August 15, 2011. As of January 31, 2012:
http://www.oecd.org/document/60/0,3746,en_2649_34533_1942460_1_1_1_1,00.html

———, "Taxation, 2004," *OECD in Figures 2007*, undated (b).

———, "Welcome to OECD.Stat Extracts," undated (c). As of May 3, 2010:
http://stats.oecd.org/Index.aspx

———, "Welcome to SourceOECD," undated (d).

———, *Mexico*, Paris, 2005.

———, *Agricultural and Fisheries Policies in Mexico: Recent Achievements, Continuing the Reform Agenda*, Paris, 2006a.

———, *Education at a Glance: OECD Indicators 2006*, Paris, 2006b.

———, *OECD Employment Outlook: Boosting Jobs and Incomes*, Paris, 2006c.

———, *Getting It Right: OECD Perspectives on Policy Challenges in Mexico*, Paris, 2007a.

———, *OECD Rural Policy Reviews: Mexico*, Paris, 2007b.

———, *A New Fiscal and Tax Policy for Mexico: A Policy Brief*, January 2007c.

———, *Growing Unequal? Income Distribution and Poverty in OECD Countries*, Paris, 2008.

———, *Economic Survey of Mexico 2011*, Paris, May 2011a.

———, *Revenue Statistics 1965–2010: 2011 Edition*, Paris, November 2011b. As of January 31, 2012:
http://www.oecd.org/document/35/0,3746,en_2649_37427_46661795_1_1_1_37427,00.html

Orozco, Manuel, *Worker Remittances: An International Comparison*, Multilateral Investment Fund, Inter-American Development Bank, working paper, February 28, 2003. As of January 31, 2012:
http://idbdocs.iadb.org/wsdocs/getdocument.aspx?docnum=35076515

Orth, Samuel P., "The Alien Contract Labor Law," *Political Science Quarterly*, Vol. 22, No. 1, March 1907, pp. 49–60.

Ottaviano, Gianmarco I. P., and Giovanni Peri, *Rethinking the Gains from Immigration: Theory and Evidence from the US*, London: Centre for Economic Policy Research, discussion paper 5226, September 2005.

———, *Immigration and National Wages: Clarifying the Theory and the Empirics*, Cambridge, Mass.: National Bureau of Economic Research, working paper 14188, 2008. As of January 31, 2012:
http://papers.nber.org/papers/w14188

PAHO—*See* Pan American Health Organization.

Palivos, Theodore, "Welfare Effects of Illegal Immigration," *Journal of Population Economics*, Vol. 22, No. 1, January 2009, pp. 131–144.

Pan American Health Organization, *Health in the Americas*, Washington, D.C., 1998.

———, *Health Situation in the Americas: Basic Indicators 2007*, Washington, D.C., 2007. As of May 3, 2010:
http://www.paho.org/english/dd/ais/BI_2007_ENG.pdf

Papademetriou, Demetrios G., and Aaron Terrazas, *Immigrants and the Current Economic Crisis: Research Evidence, Policy Challenges, and Implications*, Migration Policy Institute, January 2009. As of August 9, 2011:
http://www.migrationpolicy.org/pubs/lmi_recessionJan09.pdf

Passel, Jeffrey S., *Estimates of the Size and Characteristics of the Undocumented Population*, Washington, D.C.: Pew Hispanic Center, March 21, 2005. As of May 3, 2010:
http://pewhispanic.org/files/reports/44.pdf

———, *The Size and Characteristics of the Unauthorized Migrant Population in the U.S.: Estimates Based on the March 2005 Current Population Survey*, Washington, D.C.: Pew Hispanic Center, March 7, 2006. As of May 3, 2010:
http://pewhispanic.org/files/reports/61.pdf

Passel, Jeffrey S., and D'Vera Cohn, *U.S. Population Projections: 2005–2050*, Pew Hispanic Center, February 11, 2008a. As of August 9, 2011:
http://pewhispanic.org/files/reports/85.pdf

———, *Trends in Unauthorized Immigration: Undocumented Inflow Now Trails Legal Inflow*, Washington, D.C.: Pew Hispanic Center, October 2, 2008b. As of May 3, 2010:
http://pewhispanic.org/files/reports/94.pdf

———, *A Portrait of Unauthorized Immigrants in the United States*, Washington, D.C.: Pew Hispanic Center, April 14, 2009. As of May 3, 2010:
http://www.lrl.state.tx.us/scanned/archive/2009/8903.pdf

———, *U.S. Unauthorized Immigration Flows Are Down Sharply Since Mid-Decade*, Pew Hispanic Center, September 1, 2010. As of August 9, 2011:
http://pewhispanic.org/files/reports/126.pdf

Passel, Jeffrey S., and Roberto Suro, *Rise, Peak and Decline: Trends in U.S. Immigration 1992–2004*, Washington, D.C.: Pew Hispanic Center, September 27, 2005. As of January 31, 2012:
http://www.pewhispanic.org/2005/08/16/
attitudes-toward-immigrants-and-immigration-policy/

Pastor, Robert A., *Migration and Development in the Caribbean: The Unexplored Connection*, Boulder, Colo.: Westview Press, 1985.

———, *Toward a North American Community: Lessons from the Old World for the New*, Washington, D.C.: Institute for International Economics, 2001.

———, ed., *The Paramount Challenge for North America: Closing the Development Gap*, Washington, D.C.: American University Center for North American Studies, March 14, 2005. As of May 3, 2010:
http://www1.american.edu/ia/cnas/pdfs/NADBank.pdf

———, "The Future of North America: Replacing a Bad Neighbor Policy," *Foreign Affairs*, Vol. 87, No. 4, July–August 2008, pp. 84–98.

PEMEX—*See* Petróleos Mexicanos.

Pérez, Mamerto, Sergio Schlesinger, and Timothy A. Wise, *The Promise and the Perils of Agricultural Trade Liberalization: Lessons from Latin America*, Washington, D.C.: Washington Office on Latin America, Global Development and Environment Institute, 2008.

Peri, Giovanni, *The Effect of Immigration on Productivity: Evidence from US States*, National Bureau of Economic Research, working paper 15507, November 2009. As of August 9, 2011:
http://www.nber.org/papers/w15507

Petróleos Mexicanos, *10 Statistical Yearbook*, c. 2010. As of August 9, 2011:
http://www.ri.pemex.com/files/content/anuario_ingles_2010_6ag.pdf

———, *2011 Anuario Estadístico*, Mexico City, México, 2011a. As of January 31, 2012:
http://www.ri.pemex.com/files/content/pemex%20Anuario_a.pdf

———, "Audited Financial Results of Petróleos Mexicanos, Subsidiary Entities and Subsidiary Companies as of December 31, 2010," May 4, 2011b. As of November 1, 2011:
http://www.ri.pemex.com/files/content/
Reporte%204Q10D%20i%20201105132.pdf

Pew Hispanic Center, "Indicators of Recent Migration Flows from Mexico," fact sheet, May 30, 2007. As of September 22, 2007:
http://pewhispanic.org/files/factsheets/33.pdf

———, "Mexican Immigrants in the United States, 2008," fact sheet, April 15, 2009. As of June 10, 2009:
http://pewhispanic.org/files/factsheets/47.pdf

———, "The Mexican-American Boom: Births Overtake Immigration," July 14, 2011. As of January 31, 2012:
http://www.pewhispanic.org/files/reports/144.pdf

Pew Research Center for the People and the Press and Pew Hispanic Center, *America's Immigration Quandary: No Consensus on Immigration Problem or Proposed Fixes*, Washington, D.C., March 30, 2006. As of June 1, 2010:
http://pewhispanic.org/files/reports/63.pdf

PHC—*See* Pew Hispanic Center.

Prante, Gerald, *Special Report: Property Tax Collections Surged with Housing Prices*, Washington, D.C.: Tax Foundation, 2006. As of February 21, 2012:
http://www.taxfoundation.org/files/sr146.pdf

PRCPP and PHC—*See* Pew Research Center for the People and the Press and Pew Hispanic Center.

Presidencia de la República, undated homepage. As of February 16, 2012:
http://www.presidencia.gob.mx/gobierno/

Preston, Julia, and Samuel Dillon, *Opening Mexico: The Making of a Democracy*, New York: Farrar, Straus and Giroux, 2004.

Price, Niko, "Mexican President-Elect Details Plan for Open Border with U.S.," Associated Press, August 17, 2000.

Psacharopoulos, George, and Harry Anthony Patrinos, eds., *Indigenous Peoples and Poverty in Latin America: An Empirical Analysis*, Washington, D.C.: World Bank, 1994.

Public Law 89-236, Immigration and Nationality Act, October 3, 1965.

Public Law 91-190, National Environmental Policy Act of 1969, January 1, 1970.

Public Law 91-604, Clean Air Act of 1970, December 31, 1970.

Public Law 99-603, Immigration Reform and Control Act, November 6, 1986.

Public Law 101-649, Immigration Act, November 29, 1990.

Public Law 104-193, Personal Responsibility and Work Opportunity Reconciliation Act, August 22, 1996.

Public Law 104-208, Illegal Immigration Reform and Immigrant Responsibility Act, September 30, 1996.

Public Law 106-346, Department of Transportation and Related Agencies Appropriations Act, 2001, October 23, 2000. As of May 6, 2010:
http://www.gpoaccess.gov/serialset/cdocuments/sd106-30/pdf/745-790.pdf

Public Law 107-56, Uniting and Strengthening America by Providing Appropriate Tools Required to Intercept and Obstruct Terrorism, October 26, 2001.

Public Law 107-296, Homeland Security Act, November 25, 2002. As of January 31, 2012:
http://www.gpo.gov/fdsys/pkg/PLAW-107publ296/content-detail.html

Public Law 109-367, Secure Fence Act of 2006, October 26, 2006. As of April 29, 2010:
http://frwebgate.access.gpo.gov/cgi-bin/
getdoc.cgi?dbname=109_cong_public_laws&docid=f:publ367.109

Public Law 111-8, Omnibus Appropriations Act, 2009, March 11, 2009. As of May 6, 2010:
http://frwebgate.access.gpo.gov/cgi-bin/
getdoc.cgi?dbname=111_cong_public_laws&docid=f:publ008.111

Quintin, Erwan, and José Joaquín López, "Mexico's Financial Vulnerability: Then and Now," *Economic Letter*, Vol. 1, No. 6, June 2006. As of April 29, 2010:
http://www.dallasfed.org/research/eclett/2006/el0606.html

Ramsey, Ben, *An Evaluation of Competition Policy in Mexico*, Washington, D.C., Johns Hopkins SAIS, August 2003.

Rasmussen Reports, "56% Want NAFTA Renegotiated, Americans Divided on Free Trade," June 20, 2008. As of December 16, 2008:
http://www.rasmussenreports.com/public_content/politics/general_politics/
june_2008/56_want_nafta_renegotiated_americans_divided_on_free_trade

Rendall, Michael S., Peter Brownell, and Sarah Kups, *Declining Return Migration from the United States to Mexico in the Late-2000s Recession*, Santa Monica, Calif.: RAND Corporation, WR-720-1, 2010. As of January 31, 2012:
http://www.rand.org/pubs/working_papers/WR720-1.html

Reyes, Jesús Heroles, "Changing Perceptions," in Andrew D. Selee, ed., *Perceptions and Misconceptions in U.S.-Mexico Relations*, Washington, D.C.: Woodrow Wilson Center for Scholars, 2005, pp. 45–51.

Riosmena, Fernando, and Douglas S. Massey, *Pathways to El Norte: Origins, Destinations, and Characteristics of Mexican Migrants to the United States*, University of Colorado at Boulder, Institute of Behavioral Science, Population Program, June 2010. As of August 9, 2011:
http://www.colorado.edu/ibs/pubs/pop/pop2010-0002.pdf

Robles Vásquez, Héctor V., and Felipe Martínez Rizo, eds., *Panorama educativo de México 2005: Indicadores del sistema educativo nacional*, México: Instituto Nacional para la Evaluación de la Educación, 2006.

Rodriguez, Richard, "The New Geography of North America," in Andrew D. Selee, ed., *Perceptions and Misconceptions in U.S.-Mexico Relations*, Washington, D.C.: Woodrow Wilson Center for Scholars, 2005, pp. 19–24.

Romalis, John, *NAFTA's and CUSFTA's Impact on International Trade*, Cambridge, Mass.: National Bureau of Economic Research, working paper 11059, January 2005. As of January 31, 2012:
http://www.nber.org/papers/w11059

Rosenblum, Marc R., "Moving Beyond the Policy of No Policy: Emigration from Mexico and Central America," *Latin American Politics and Society*, Vol. 46, No. 4, Winter 2004, pp. 91–126.

———, *Immigration Enforcement at the Worksite: Making It Work*, Migration Policy Institute, policy brief 6, November 2005. As of August 9, 2011:
http://www.migrationpolicy.org/ITFIAF/PolicyBrief-6-Rosenblum.pdf

———, "The United States and Mexico: Prospects for a Bilateral Migration Policy," *Border Battles: The U.S. Immigration Debates*, March 8, 2007. As of May 3, 2010:
http://borderbattles.ssrc.org/Rosenblum/

Ruggles, Steven, J. Trent Alexander, Katie Genadek, Ronald Goeken, Matthew B. Schroeder, and Matthew Sobek, *Integrated Public Use Microdata Series*, version 5.0, Minneapolis, Minn.: University of Minnesota, 2010.

S.1033—*See* U.S. Senate (2005a).

S.1348—*See* U.S. Senate (2007).

S.1438—*See* U.S. Senate (2005b).

S.1645—*See* U.S. Senate (2003).

S.2611—*See* U.S. Senate (2006).

Sánchez Daza, Alfredo, "Transformación financiera de México," in García Alba Iduñate, Lucino Gutiérrez Herrera Pascual, and Gabriela Torres Ramírez, eds., *El nuevo milenio Mexicano*, Tomo 2: *Economía, ahorro y finanzas*, Universidad Autónoma Metropolitana, Unidad Azcapotzalco: Ediciones y Gráficos Eón, 2004, pp. 233–269.

Santibañez, Lucrecia, Georges Vernez, and Paula Razquin, *Education in Mexico: Challenges and Opportunities*, Santa Monica, Calif.: RAND Corporation, DB-480-HF, 2005. As of January 31, 2012:
http://www.rand.org/pubs/documented_briefings/DB480.html

Schaefer, Agnes Gereben, Benjamin Bahney, and K. Jack Riley, *Security in Mexico: Implications for U.S. Policy Options*, Santa Monica, Calif.: RAND Corporation, MG-876-RC, 2009. As of January 31, 2012:
http://www.rand.org/pubs/monographs/MG876.html

Schiller, Nina Glick, Linda Basch, and Cristina Szanton Blanc, "From Immigrant to Transmigrant: Theorizing Transnational Migration," *Anthropological Quarterly*, Vol. 68, No. 1, January 1995, pp. 48–63.

Scovel, Calvin L. III, inspector general, U.S. Department of Transportation, "Cross-Border Trucking Demonstration Project," statement before the Committee on Commerce, Science, and Transportation, U.S. Senate, March 11, 2008.

Secretaría de Desarrollo Social, "Programa Nacional de Desarrollo Social, 2001–2006," Mexico City, 2001.

Secretaría de Economía, "Appris suscritos," undated (a).

———, "Sector externo: Inversión extranjera directa: Por país de origen," Dirección General de Inversión Extranjera, undated (b).

———, "Estado de los appris suscritos por México," 2007a.

———, "Importaciones totales de México and exportaciones totales de México," Inteligencia Comercial, Subsecretaría de Negociaciones Comerciales Internacionales, 2007b.

———, "Trade Agreements," 2007c.

———, "Informe estadístico trimestral sobre el comportamiento de la inversión extranjera directa en México," February 26, 2010.

———, "Mexico's Total Exports: Value in Million Dollars," Underministry of Foreign Trade, c. 2011a. As of August 9, 2011:
http://www.economia-snci.gob.mx/sic_php/pages/estadisticas/cuad_resumen/expmx_i.pdf

———, "Mexico's Total Imports: Value in Million Dollars," Underministry of Foreign Trade, c. 2011b. As of August 9, 2011:
http://www.economia-snci.gob.mx/sic_php/pages/estadisticas/cuad_resumen/impmx_i.pdf

Secretaría de Educación Pública, "Sistemas para el análisis de la estadística educativa," undated web page. As of May 4, 2010:
http://dgpp.sep.gob.mx/Estadi/SistesepPortal/sistesep.html

Secretaría de Hacienda y Crédito Público, "Public-Sector Budgetary Revenues," undated.

———, *Plan anual de financiamiento*, 2011. As of February 16, 2012:
http://www.shcp.gob.mx/POLITICAFINANCIERA/UCP/PCP/Plan%20Anual%20de%20Financiamiento/PAF%202011.pdf

Sedano, Fernando, "Economic Implications of Mexico's Sudden Demographic Transition," *Business Economics*, Vol. 43, No. 3, July 2008, pp. 40–54.

SEDESOL—*See* Secretaría de Desarrollo Social.

Selee, Andrew, and Katie Putnam, "Mexico's 2009 Midterm Elections: Winners and Losers," Mexico Institute, Woodrow Wilson International Center for Scholars, July 2009.

SEP—*See* Secretaría de Educación Pública.

Shaffer, Jay C., *Competition Law and Policy in Mexico: An OECD Peer Review*, Paris: Organisation for Economic Co-Operation and Development, 2004. As of January 31, 2012:
http://www.oecd.org/dataoecd/57/9/31430869.pdf

Shah, Anwar, *Fiscal Decentralization in Developing and Transition Economies: Progress, Problems, and the Promise*, Washington, D.C.: World Bank, working paper 3282, 2004.

Shatz, Howard Jerome, *The Location of U.S. Multinational Affiliates*, Cambridge, Mass.: Harvard University, Ph.D. thesis, May 2000.

Shatz, Howard Jerome, and Louis Felipe López-Calva, *The Emerging Integration of the California-Mexico Economies*, San Francisco, Calif.: Public Policy Institute of California, 2004.

SHCP—*See* Secretaría de Hacienda y Crédito Público.

SINAIS—*See* Sistema Nacional de Información en Salud.

Sistema Nacional de Información en Salud, "Mortalidad: Información tabular," last modified November 3, 2008. As of May 4, 2010:
http://sinais.salud.gob.mx/mortalidad/

Smith, James P., and Barry Edmonston, eds., *The New Americans: Economic, Demographic, and Fiscal Effects of Immigration*, Washington, D.C.: National Academy Press, 1997.

Sobarzo, Horacio, *Tax Effort and Tax Potential of State Governments in Mexico: A Representative Tax System*, Notre Dame, Ind.: Helen Kellogg Institute for International Studies, working paper 315, October 2004. As of January 31, 2012:
http://kellogg.nd.edu/publications/workingpapers/WPS/315.pdf

Social Security Administration, "United States/Mexico Totalization Agreement," June 2004. As of January 31, 2012:
http://www.ssa.gov/pressoffice/factsheets/USandMexico.htm

Soto Priante, Sergio, and Marco Antonio Velázquez Holguín, "El proceso de institucionalización del Programa 3x1 para migrantes," in Rafael Fernández de Castro, Rodolfo García Zamora, and Ana Vila Freyer, eds., *El Programa 3x1 para migrantes: ¿Primera política transnacional en México?* México: Instituto Tecnológico Autónomo de México, Universidad Autónoma de Zacatecas, 2006.

Study Group on U.S.-Mexico Relations, *The United States and Mexico: Forging a Strategic Partnership*, Washington, D.C.: Woodrow Wilson International Center for Scholars Mexico Institute, 2005.

Swiss National Bank, "Prices and Salaries/Wages: O3 Prices of Precious Metals and Raw Materials," *Monthly Statistical Bulletin*, July 2011, p. 116. As of January 31, 2012:
http://www.snb.ch/en/iabout/stat/statpub/statmon/stats/statmon/statmon_O3

Tanner, Jeffery, *Smart Choices? Cognitive Impacts of Children Left Behind by Paternal International Migration*, unpublished, prepared for Midwest International Economic Development Conference, April 2010.

Ter-Minassian, Teresa, *Fiscal Federalism in Theory and Practice*, Washington, D.C.: International Monetary Fund, 1997.

Terrazas, Aaron, "Mexican Immigrants in the United States," *Migration Information Source*, February 2010. As of August 9, 2011:
http://www.migrationinformation.org/USfocus/display.cfm?id=767

Tuirán Gutiérrez, Rodolfo, Jorge Santibáñez Romellón, and Rodolfo Corona Vázquez, "El monto de las remesas familiares en México: ¿Mito o realidad?" *Papeles de Población*, No. 50, October–December 2006, pp. 147–169.

UNDP—*See* United Nations Development Programme.

UNESCO—*See* United Nations Educational, Scientific and Cultural Organization.

United Nations Development Programme, *Informe Sobre Desarrollo Humano México 2006–2007*, 2007.

———, *Human Development Report 2010: 20th Anniversary Edition—The Real Wealth of Nations: Pathways to Human Development*, 2010. As of February 16, 2012:
http://hdr.undp.org/en/reports/global/hdr2010/

United Nations Educational, Scientific and Cultural Organization, *Primer estudio internacional comparativo sobre lenguaje, matemática y factores asociados, para alumnos del tercer y cuarto grado de la educación básica: Informe técnico*, Santiago, Chile, 1998. As of May 4, 2010:
http://portal.unesco.org/geography/es/
ev.php-URL_ID=8308&URL_DO=DO_TOPIC&URL_SECTION=201.html

United Nations Population Division, *World Population Prospects: The 2006 Revision Population Database*, March 13, 2007. As of January 31, 2012:
http://www.un.org/esa/population/publications/wpp2006/wpp2006.htm

UNPD—*See* United Nations Population Division.

U.S. Census Bureau, "International Data Base," undated (a). As of August 9, 2011:
http://www.census.gov/population/international/data/idb/
informationGateway.php

———, "Place of Birth for the Foreign-Born Population [126]—Universe: Foreign-Born Population," table PCT19 of Census 2000 Summary File 3 sample data, undated (b).

———, *Current Population Survey*, March 1994.

———, *Current Population Survey*, March 1995.

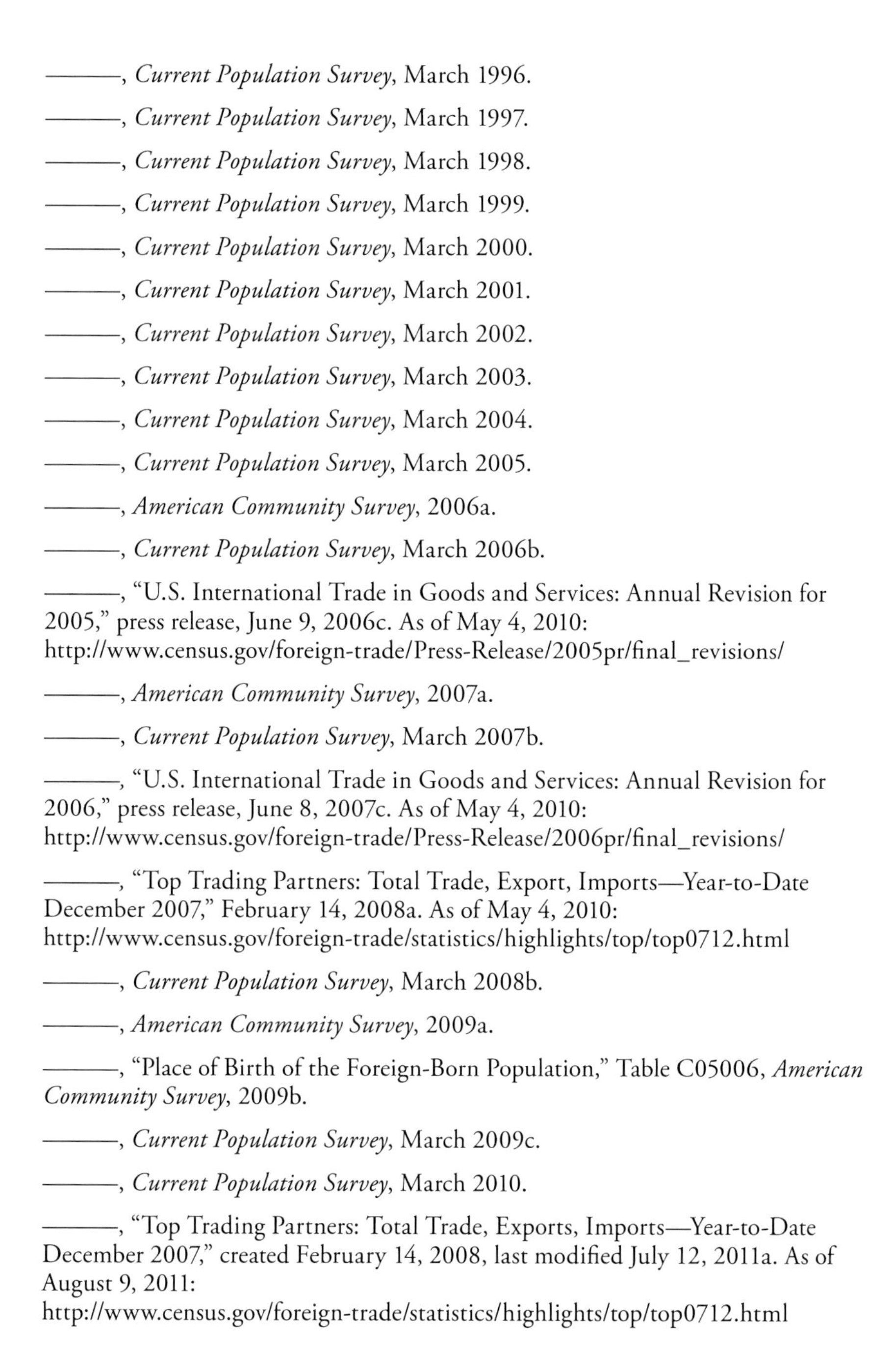

———, *Current Population Survey*, March 1996.

———, *Current Population Survey*, March 1997.

———, *Current Population Survey*, March 1998.

———, *Current Population Survey*, March 1999.

———, *Current Population Survey*, March 2000.

———, *Current Population Survey*, March 2001.

———, *Current Population Survey*, March 2002.

———, *Current Population Survey*, March 2003.

———, *Current Population Survey*, March 2004.

———, *Current Population Survey*, March 2005.

———, *American Community Survey*, 2006a.

———, *Current Population Survey*, March 2006b.

———, "U.S. International Trade in Goods and Services: Annual Revision for 2005," press release, June 9, 2006c. As of May 4, 2010:
http://www.census.gov/foreign-trade/Press-Release/2005pr/final_revisions/

———, *American Community Survey*, 2007a.

———, *Current Population Survey*, March 2007b.

———, "U.S. International Trade in Goods and Services: Annual Revision for 2006," press release, June 8, 2007c. As of May 4, 2010:
http://www.census.gov/foreign-trade/Press-Release/2006pr/final_revisions/

———, "Top Trading Partners: Total Trade, Export, Imports—Year-to-Date December 2007," February 14, 2008a. As of May 4, 2010:
http://www.census.gov/foreign-trade/statistics/highlights/top/top0712.html

———, *Current Population Survey*, March 2008b.

———, *American Community Survey*, 2009a.

———, "Place of Birth of the Foreign-Born Population," Table C05006, *American Community Survey*, 2009b.

———, *Current Population Survey*, March 2009c.

———, *Current Population Survey*, March 2010.

———, "Top Trading Partners: Total Trade, Exports, Imports—Year-to-Date December 2007," created February 14, 2008, last modified July 12, 2011a. As of August 9, 2011:
http://www.census.gov/foreign-trade/statistics/highlights/top/top0712.html

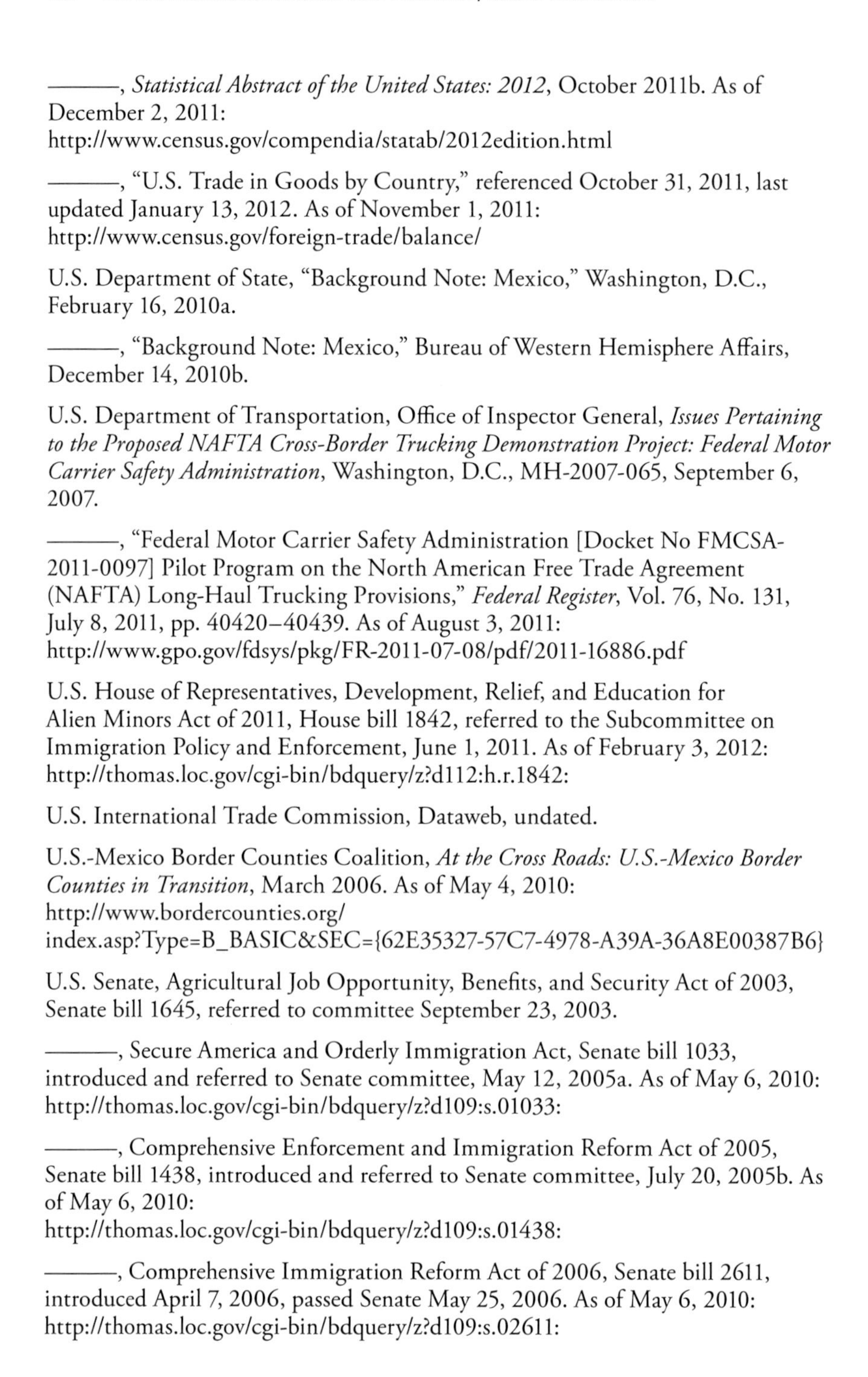

———, *Statistical Abstract of the United States: 2012*, October 2011b. As of December 2, 2011:
http://www.census.gov/compendia/statab/2012edition.html

———, "U.S. Trade in Goods by Country," referenced October 31, 2011, last updated January 13, 2012. As of November 1, 2011:
http://www.census.gov/foreign-trade/balance/

U.S. Department of State, "Background Note: Mexico," Washington, D.C., February 16, 2010a.

———, "Background Note: Mexico," Bureau of Western Hemisphere Affairs, December 14, 2010b.

U.S. Department of Transportation, Office of Inspector General, *Issues Pertaining to the Proposed NAFTA Cross-Border Trucking Demonstration Project: Federal Motor Carrier Safety Administration*, Washington, D.C., MH-2007-065, September 6, 2007.

———, "Federal Motor Carrier Safety Administration [Docket No FMCSA-2011-0097] Pilot Program on the North American Free Trade Agreement (NAFTA) Long-Haul Trucking Provisions," *Federal Register*, Vol. 76, No. 131, July 8, 2011, pp. 40420–40439. As of August 3, 2011:
http://www.gpo.gov/fdsys/pkg/FR-2011-07-08/pdf/2011-16886.pdf

U.S. House of Representatives, Development, Relief, and Education for Alien Minors Act of 2011, House bill 1842, referred to the Subcommittee on Immigration Policy and Enforcement, June 1, 2011. As of February 3, 2012:
http://thomas.loc.gov/cgi-bin/bdquery/z?d112:h.r.1842:

U.S. International Trade Commission, Dataweb, undated.

U.S.-Mexico Border Counties Coalition, *At the Cross Roads: U.S.-Mexico Border Counties in Transition*, March 2006. As of May 4, 2010:
http://www.bordercounties.org/
index.asp?Type=B_BASIC&SEC={62E35327-57C7-4978-A39A-36A8E00387B6}

U.S. Senate, Agricultural Job Opportunity, Benefits, and Security Act of 2003, Senate bill 1645, referred to committee September 23, 2003.

———, Secure America and Orderly Immigration Act, Senate bill 1033, introduced and referred to Senate committee, May 12, 2005a. As of May 6, 2010:
http://thomas.loc.gov/cgi-bin/bdquery/z?d109:s.01033:

———, Comprehensive Enforcement and Immigration Reform Act of 2005, Senate bill 1438, introduced and referred to Senate committee, July 20, 2005b. As of May 6, 2010:
http://thomas.loc.gov/cgi-bin/bdquery/z?d109:s.01438:

———, Comprehensive Immigration Reform Act of 2006, Senate bill 2611, introduced April 7, 2006, passed Senate May 25, 2006. As of May 6, 2010:
http://thomas.loc.gov/cgi-bin/bdquery/z?d109:s.02611:

———, Comprehensive Immigration Reform Act of 2007, Senate bill 1348, introduced May 9, 2007. As of May 6, 2010:
http://thomas.loc.gov/cgi-bin/bdquery/z?d110:s.01348:

———, Development, Relief, and Education for Alien Minors Act of 2010, Senate bill 3992, cloture motion on the motion to proceed to the bill rendered moot in Senate, December 9, 2010. As of February 3, 2012:
http://thomas.loc.gov/cgi-bin/bdquery/z?d111:SN03992:

———, Development, Relief, and Education for Alien Minors Act of 2011, Senate bill 952, hearings held in the Committee on the Judiciary Subcommittee on Immigration, Refugees and Border Security, June 28, 2011. As of February 3, 2012:
http://thomas.loc.gov/cgi-bin/bdquery/z?d112:SN00952:

Van't Hek, Koen, Michael Becka, and Rocio Mejia, "Mexican Government Enacts Important Changes to IMMEX (Formerly Maquiladora) Regime," Ernst and Young, undated. As of August 9, 2011:
http://www.ey.com/Publication/vwLUAssets/IMMEX_Newsletter/$FILE/comentarios_832.pdf

Waslin, Michelle, *The New Meaning of the Border: U.S.-Mexico Migration Since 9/11*, Washington, D.C.: National Council of La Raza, prepared for the conference on Reforming the Administration of Justice in Mexico at the Center for U.S.-Mexican Studies, May 15–17, 2003.

WDI—*See* World Bank (undated).

Weingast, Barry R., *The Performance and Stability of Federalism: An Institutional Perspective*, Forum Series on the Role of Institutions in Promoting Economic Growth, Forum 7: Institutional Barriers to Economic Change: Cases Considered, June 24, 2003. As of June 1, 2010:
http://www.usaid.gov/our_work/economic_growth_and_trade/eg/forum_series/f7_weingast.pdf

White House, "Fact Sheet: Fair and Secure Immigration Reform," press release January 7, 2004.

———, "The Security and Prosperity Partnership of North America: Progress," press release, March 31, 2006.

WHO—*See* World Health Organization.

World Bank, "Country and Lending Groups," undated (a). As of February 16, 2012:
http://data.worldbank.org/about/country-classifications/country-and-lending-groups

———, "Doing Business Data," undated (b). As of February 16, 2012:
http://www.doingbusiness.org/data

———, World Development Indicators, Washington, D.C., undated (c).

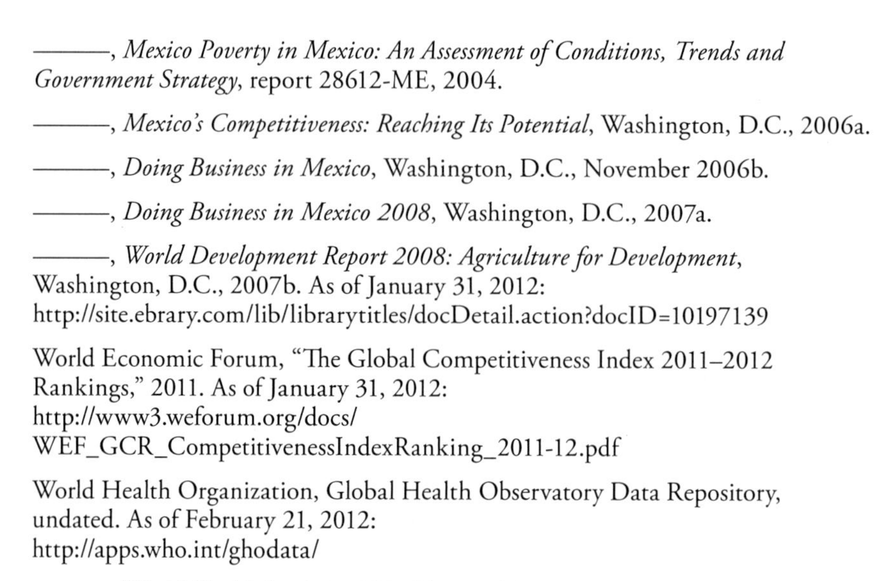

———, *Mexico Poverty in Mexico: An Assessment of Conditions, Trends and Government Strategy*, report 28612-ME, 2004.

———, *Mexico's Competitiveness: Reaching Its Potential*, Washington, D.C., 2006a.

———, *Doing Business in Mexico*, Washington, D.C., November 2006b.

———, *Doing Business in Mexico 2008*, Washington, D.C., 2007a.

———, *World Development Report 2008: Agriculture for Development*, Washington, D.C., 2007b. As of January 31, 2012:
http://site.ebrary.com/lib/librarytitles/docDetail.action?docID=10197139

World Economic Forum, "The Global Competitiveness Index 2011–2012 Rankings," 2011. As of January 31, 2012:
http://www3.weforum.org/docs/
WEF_GCR_CompetitivenessIndexRanking_2011-12.pdf

World Health Organization, Global Health Observatory Data Repository, undated. As of February 21, 2012:
http://apps.who.int/ghodata/

———, *World Health Statistics 2007*, France, 2007.

Yúnez-Naude, Antonio, and Fernando Barceinas Paredes, *The Agriculture of Mexico After Ten Years of NAFTA Implementation*, Santiago: Central Bank of Chile, working paper 277, December 2004. As of January 31, 2012:
http://www.bcentral.cl/eng/studies/working-papers/pdf/dtbc277.pdf

Yúnez-Naude, Antonio, Fernando Barceinas, and Gabriela Soto Ruiz, "El campo Mexicano en los albores del siglo XXI," in Pascual García Alba Iduñate, Lucino Gutiérrez Herrera, and Gabriela Torres Ramírez, eds., *El nuevo milenio Mexicano*, México: Universidad Autónoma Metropolitana, Unidad Azcapotzalco: Ediciones y Gráficos Eón, 2004, pp. 183–213.

Zárate-Hoyos, Germán, "The Development Impact of Migrant Remittances in México," in Donald F. Terry and Steven R. Wilson, eds., *Beyond Small Change: Making Migrant Remittances Count*, Washington, D.C.: Inter-American Development Bank, 2005.

Zúñiga Herrera, Elena, Paula Leite, and Alma Rosa Nava, *La nueva era de las migraciones: Características de la migración internacional en México*, México, D.F.: Consejo Nacional de Población, 2004.